Remote Sensing of Impervious Surfaces

in Tropical and Subtropical Areas

Taylor & Francis Series in
Remote Sensing Applications

Series Editor
Qihao Weng
Indiana State University
Terre Haute, Indiana, U.S.A.

Remote Sensing of Impervious Surfaces

in Tropical and Subtropical Areas

Hongsheng Zhang
Hui Lin
Yuanzhi Zhang
Qihao Weng

CRC Press
Taylor & Francis Group
Boca Raton London New York

CRC Press is an imprint of the
Taylor & Francis Group, an **informa** business

CRC Press
Taylor & Francis Group
6000 Broken Sound Parkway NW, Suite 300
Boca Raton, FL 33487-2742

First issued in paperback 2019

ISBN-13: 978-1-4822-5483-9 (hbk)
ISBN-13: 978-0-367-87062-1 (pbk)

Visit the Taylor & Francis Web site at
http://www.taylorandfrancis.com

and the CRC Press Web site at
http://www.crcpress.com

Contents

List of Figures

List of Tables

Preface

Remote Sensing for Urbanization in Tropical and Subtropical Regions—Why and What Matters?

The twenty-first century is the first "urban century" according to the United Nations Development Program. Although urbanized land currently covers only approximately 2% of global land area, more than half of the world's population live in the urban environment. By 2030, urbanized areas will expand to provide homes for 81% of the world's population, with the majority of the population increase coming from developing countries. Thus, there is a rapidly growing need for technologies that will allow for the monitoring of the world's urban assets and management of the exposure to natural and man-made risks. This need is further driven by increased concern over global climate change. Geographically, most developed countries are located in temperate regions, whereas developing countries are located in the tropical and subtropical regions. The continued urbanization in the tropical and subtropical regions has an important implication in biodiversity, the well-being of tropical rainforest ecosystem, and global climate change.

A characteristic change associated with urbanization is the expansion of impervious surface. Satellite remote sensing provides the only viable option to detect and monitor impervious surface from space in an efficient, affordable, and timely manner. Numerous previous studies have utilized satellite imagery of different spatial resolutions to estimate and map impervious surface (Weng 2012). The nightlight derived from the DMSP OLS and MODIS land cover products was used to produce the global urban dataset at 1 km resolution (Elvidge et al. 1997; Imhoff et al. 1997; Friedl et al. 2002; Schneider et al. 2003; Zhou et al. 2014). Global urban maps at coarse resolution can cover large areas and also be updated frequently. However, due to the complexity of urban landscapes and inherent resolution of human activities, coarse-resolution global urban maps are difficult to use for many applications at local to regional scales (Small 2003). Medium-resolution satellite imagery possesses unique advantages in mapping urban areas more accurately. Sensors on board the Landsat series of satellites have been providing Earth observation data continuously since the early 1970s (Townshend et al. 1991; Loveland and Shaw 1996), which have been applied in numerous urbanization studies at the local, regional, and continental scales (Seto et al. 2002; Jantz et al. 2005; Schneider et al. 2005). As part of the National Land Cover

Dataset, impervious surface maps were produced for the United States from Landsat data for 2006 and 2011 (Xian et al. 2011). However, the majority of previous urban land cover and land use studies tended to focus on a single image at one time (Schneider et al. 2003). Applications using Landsat data are surging due to the availability of free Landsat data from the US Geological Survey since 2008 (Woodcock et al. 2008). New methods and techniques are being developed to utilize abundant medium-resolution images and produce consistent maps for monitoring urban expansion (Sexton et al. 2013; Zhu and Woodcock 2014). The Landsat time-series data will also allow for more detailed studies to determine the impacts of urbanization on energy, water, carbon cycles, vegetation phenology, and surface climate (Weng and Fu 2014).

Continuity of medium-resolution data is critical for monitoring land use and land cover changes worldwide. However, the failure of the Scan-Line Corrector on board the Landsat 7 satellite in 2003 caused a loss of 25% of the data toward the edges of each image; Landsat 5 suspended operations in November 2011. Although Landsat-8 OLI data was available after 2013, maintaining the continuity of Landsat-like data is precarious. This situation highlights the need to combine the capabilities of existing and future international sensors to provide a more robust observational record (Weng et al. 2014). When considering international satellite missions such as Sentinel, CBERS-2, and IRS, the rich source of medium-resolution remotely sensed data suggests that we may now move urban mapping from the local and regional, to the global scale. Despite the great potential for the combined use of existing and future medium-resolution imagery, many issues deserve to be studied further, including cross-sensor comparison and normalization (Schroeder et al. 2006; Wulder et al. 2008), multisensor fusion (Gao et al. 2006; Weng et al. 2014), and utilization of full suite of Landsat-like data for any location and date (Powell et al. 2007; Gao et al. 2012). Significant challenges remain for mapping urbanization over large areas, in terms of validation and systematically processing data from multiple times, various sources/instruments, and different seasons (Gao et al. 2012).

In the tropical and subtropical regions, remote sensing of urban environment faces more challenges than in the temperate zones due to all-year-round cloudy and rainy climate conditions, complex hydrological systems that often display a strong seasonal change in water surface area, and vegetation phenology and morphological and species complexity. Optical data frequently show their weakness in remote sensing in the tropical and subtropical regions, which prompts researchers to use different sources of imagery from microwave remote sensing. Synthetic aperture radar (SAR), for instance, was widely employed previously to provide complementary information to optical imagery because it works on all-weather conditions, free from the influence of clouds and rains. Previous studies show that SAR is very sensitive to ground surface roughness, shape, structure, and dielectric properties of illuminated ground targets (Henderson and Xia 1997).

However, more research is needed to understand the methods of feature extraction, selection of fusion level, and classification algorithm when optical and SAR image data are combined to estimate and map impervious surface (Jiang et al. 2009; Yang et al. 2009; Zhang et al. 2012, 2014).

In December 2013, Dr. Hongsheng Zhang from the Chinese University of Hong Kong contacted me and expressed an interest in publishing an authored book in the Taylor & Francis Group's Remote Sensing Applications Series. Upon reading the proposal, I fully agreed on the scope and objectives as well as the table of contents. I served in Dr. Zhang's dissertation committee and have been closely watching his professional growth over the past several years. I am amazed by his aggression in research and his production of peer-refereed articles. The outlets of his publications include nearly all reputed remote sensing journals, and his work has been funded by the Hong Kong Research Council and the China Natural Science Foundation. I am pleased to have such an opportunity to introduce his work to the readers of CRC Press. By carefully selecting several case studies in different continents, this book illustrates various recent methods in the synergistic use of optical and SAR data for estimating and mapping of impervious surface for the tropical and subtropical cities and answers two fundamental science questions: (1) why SAR data are able to improve the accuracy of the estimation, and (2) how to optimize the fusion between optical and SAR data in order to achieve accurate estimation. The findings of this research shed important light into other aspects of urban as well as environmental remote sensing.

Qihao Weng
Indiana State University, Terre Haute, Indiana, USA

References

Elvidge, C. D., K. E. Baugh, E. A. Kihn, H. W. Kroehl, and E. R. Davis. 1997. Mapping city lights with nighttime data from the DMSP Operational Linescan System. *Photogrammetric Engineering & Remote Sensing* 63(6):727–734.

Friedl, M. A., D. K. McIver, J. C. F. Hodges, X. Zhang, D. Muchoney, A. H. Strahler, C. E. Woodcock, S. Gopal, A. Schnieder, A. Cooper, A. Baccini, F. Gao, and C. B. Schaaf. 2002. Global land cover from MODIS: Algorithms and early results. *Remote Sensing of Environment* 83:287–302.

Gao, F., E. B. de Colstoun, R. Ma, Q. Weng, J. G. Masek, J. Chen, Y. Pan, and C. Song. 2012. Mapping impervious surface expansion using medium-resolution satellite image time series: A case study in the Yangtze River Delta, China. *International Journal of Remote Sensing* 33(24):7609–7628.

Gao, F., J. Masek, M. Schwaller, and H. Forrest. 2006. On the blending of the Landsat and MODIS surface reflectance: Predict daily Landsat surface reflectance. *IEEE Transactions on Geoscience and Remote Sensing* 44(8):2207–2218.

Henderson, F. M., and Z. G. Xia. 1997. SAR applications in human settlement detection, population estimation and urban land use pattern analysis: A status report. *IEEE Transactions on Geoscience and Remote Sensing* 35:79–85.

Imhoff, M. L., W. T. Lawrence, D. C. Stutzer, and C. D. Elvidge. 1997. A technique for using composite DMSP/OLS "city lights" satellite data to map urban areas. *Remote Sensing of Environment* 61:361–370.

Jantz, P., S. Goetz, and C. Jantz. 2005. Urbanization and the loss of resource lands in the Chesapeake Bay Watershed. *Environmental Management* 36:808–825.

Jiang, L. M., M. S. Liao, H. Lin, H., and L. M. Yang. 2009. Synergistic use of optical and InSAR data for urban impervious surface mapping: A case study in Hong Kong. *International Journal of Remote Sensing* 30:2781–2796.

Loveland, T. R., and D. M. Shaw. 1996. Multi-resolution land characterization: Building collaborative partnerships, in GAP Analysis: A Landscape Approach to Biodiversity Planning. In: *American Society for Photogrammetry and Remote Sensing*, J. M. Scott, T. H. Tear, and F. W. Davis, eds., 79–85. Bethesda, Maryland.

Powell, S. L., D. Pflugmacher, A. A. Kirschbaum, Y. Kim, and W. B. Cohen. 2007. Moderate resolution remote sensing alternatives: A review of Landsat-like sensors and their applications. *Journal of Applied Remote Sensing* 1:012506.

Schneider, A., M. A. Friedl, D. K. McIver, and C. E. Woodcock. 2003. Mapping urban areas by fusing multiple sources of coarse resolution remotely sensed data. *Photogrammetric Engineering & Remote Sensing* 69:1377–1386.

Schneider, A., K. C. Seto, and D. R. Webster. 2005. Urban growth in Chengdu, Western China: Application of remote sensing to assess planning and policy outcomes. *Environment and Planning B-Planning & Design* 32(3):323–345.

Schroeder, T. A., W. B. Cohen, C. Song, M. J. Canty, and Z. Yang. 2006. Radiometric correction of multi-temporal Landsat data for characterization of early successional forest patterns in western Oregon. *Remote Sensing of Environment* 103:16–26.

Seto, K. C., C. E. Woodcock, C. Song, X. Huang, R. K. Kaufmann, and J. Lu. 2002. Measuring landuse change with Landsat TM: Evidence from Pearl River Delta. *International Journal of Remote Sensing* 23:1985–2004.

Sexton, J. O., X.-P. Song, C. Huang, S. Channan, M. E. Baker, and J. R. Townshend. 2013. Urban growth of the Washington, D.C.–Baltimore, MD metropolitan region from 1984 to 2010 by annual, Landsat-based estimates of impervious cover. *Remote Sensing of Environment* 129:42–53.

Small, C. 2003. High spatial resolution spectral mixture analysis of urban reflectance. *Remote Sensing of Environment* 88:170–186.

Townshend, J., C. Justice, W. Li, C. Gurney, and J. McManus. 1991. Global land cover classification by remote sensing: Present capabilities and future possibilities. *Remote Sensing of Environment* 35:243–255.

Weng, Q. 2012. Remote sensing of impervious surfaces in the urban areas: Requirements, methods, and trends. *Remote Sensing of Environment* 117(2):34–49.

Weng, Q., and P. Fu. 2014. Modeling annual parameters of land surface temperature variations and evaluating the impact of cloud cover using time series of Landsat TIR data. *Remote Sensing of Environment* 140:267–278.

Weng, Q., P. Fu, and F. Gao. 2014. Generating daily land surface temperature at Landsat resolution by fusing Landsat and MODIS data. *Remote Sensing of Environment* 145:55–67.

Weng, Q., T. Esch, P. Gamba, D. A. Quattrochi, and G. Xian. 2014. Global urban observation and information: GEO's effort to address the impacts of human settlements. In Weng, Q. ed. *Global Urban Monitoring and Assessment through Earth Observation*, Chapter 2, 15–34. Boca Raton, FL: CRC Press/Taylor and Francis.

Woodcock, C. E., R. Allen, M. Anderson, A. Belward, R. Bindschadler, W. Cohen, F. Gao, S. N. Goward, D. Helder, E. Helmer, R. Nemani, L. Oreopoulos, J. Schott, P. S. Thenkabail, E. F. Vermote, J. Vogelmann, M. A. Wulder, and R. Wynne. 2008. Free access to Landsat imagery. *Science* 320:1011–1011.

Wulder, M. A., C. R. Butson, and J. C. White. 2008. Cross-sensor change detection over a forested landscape: Options to enable continuity of medium spatial resolution measures. *Remote Sensing of Environment* 112(3):796–809.

Xian, G., C. Homer, J. Dewitz, J. Fry, N. Hossain, and J. Wickham. 2011. The change of impervious surface area between 2001 and 2006 in the conterminous United States. *Photogrammetric Engineering & Remote Sensing* 77(8):758–762.

Yang, L. M., L. M. Jiang, H. Lin, and M. S. Liao. 2009. Quantifying sub-pixel urban impervious surface through fusion of optical and InSAR imagery. *GIScience & Remote Sensing* 46:161–171.

Zhang, H. S., Y. Zhang, and H. Lin. 2012. A comparison study of impervious surfaces estimation using optical and SAR remote sensing images. *International Journal of Applied Earth Observation and Geoinformation* 18:148–156.

Zhang, Y., H. S. Zhang, and H. Lin. 2014. Improving the impervious surface estimation with combined use of optical and SAR remote sensing images. *Remote Sensing of Environment* 141:155–167.

Zhou, Y., S. J. Smith, C. D. Elvidge, K. Zhao, A. Thomson, and M. Imhoff. 2014. A cluster-based method to map urban area from DMSP/OLS nightlights. *Remote Sensing of Environment* 147:173–185.

Zhu, Z., and C. E. Woodcock. 2014. Automated cloud, cloud shadow, and snow detection in multitemporal Landsat data: An algorithm designed specifically for monitoring land cover change. *Remote Sensing of Environment* 152:217–234.

Acknowledgments

This study is mainly supported by the National Natural Science Foundation of China (41401370). It is jointly supported by the National Basic Research Program of China (2015CB954103), the General Research Fund (CUHK 457212), and the National Natural Science Foundation of China (41171288). The University of Pavia and the German Aerospace Agency (DLR) are greatly appreciated for providing the TerraSAR-X data. Landsat TM/ETM+ data provided by the US Geological Survey are also greatly appreciated. The authors also thank the Institute of Space and Earth Information Science (ISEIS)/The Chinese University of Hong Kong (CUHK) for providing ENVISAT ASAR data. Special thanks to colleagues from the University of Pavia, Indiana State University, ISEIS/CUHK, and the Shenzhen Research Institute/CUHK for their valuable discussion and comments toward our study. The authors also thank three anonymous reviewers for providing critical comments and suggestions that have improved the original manuscript.

List of Abbreviations

ANN	artificial neural network
BIS	bright impervious surface
BP	backpropagating
BS	bare soil
CART	classification and regression tree
CF	color feature
DEM	digital elevation model
DGPS	differential GPS
DIS	dark impervious surface
DISS	dissimilarity
DOP	digital orthophoto
DT	decision tree
ENT	entropy
ERM	empirical risk minimization
FOV	field of view
GCP	ground control point
GLCM	gray-level co-occurrence matrix
HOM	homogeneity
HSI	hue-saturation-intensity
HSR	humid subtropical region
HSV	hue-saturation-value
ISE	impervious surface estimation
ISODATA	iterative self-organizing data analysis technique algorithm
Kappa	Kappa coefficient
LULC	land use/land cover
MLC	maximum likelihood classifier
MLP	multilayer perceptron
MNF	maximum noise fraction
NDISI	normalized difference impervious surfaces index
NDVI	normalized difference vegetation index
NDWI	normalized difference water index
NPS	nonpoint source
NSMA	normalized spectral mixture analysis
NVT	Neuromorphic Vision Toolkit
OA	overall accuracy
OOB	out-of-bag
PRD	Pearl River Delta
PRE	Pearl River Estuary
RF	random forest
RGB	red-green-blue

RMSE	root-mean-square error
RS	remote sensing
SAN	shape-adaptive neighbourhood
SAR	synthetic aperture radar
SEZ	special economic zone
SF	spectral feature
SLC	scan line corrector
SMA	spectral mixture analysis
SRM	structural risk minimization
SVM	support vector machine
TF	texture feature
TSX	TerraSAR-X
UHI	urban heat island
UTM	Universal Transverse Mercator
VIS	vegetation-impervious surface-soil
VSA	variable source areas
WGS84	World Geodetic System 1984

1

Introduction

1.1 Research Background

Dramatic urbanization processes have occurred in many regions around the world and thus have created a number of metropolises, especially in tropical and subtropical regions. One of the most important implications of this occurrence is that a large portion of impervious surface is the result of this rapid urbanization process. Impervious surfaces have been widely recognized as the most important land cover type in urban areas, and they serve as a key environmental indicator of many environmental issues such as urban flooding, urban climate, water pollution, and air pollution (Arnold and Gibbons 1996; Hu and Weng 2011). Moreover, impervious surfaces are also reported to be a significant factor in many socioeconomic studies, including urban growth, estimation of population distribution, and variation of housing prices (Wu and Murray 2003; Wu and Yuan 2007; Yang et al. 2003a; Yu and Wu 2004).

Most developed countries and cities are located in temperate regions, and therefore most previous studies about impervious surface estimation (ISE) using remote sensing are focused on temperate urbanized areas. However, many developing countries, such as China and India, have been undergoing dramatic urbanization processes in the past decades. Moreover, unlike in developed countries, urbanization planning and management in these newly developed cities are not as advanced as those in developed countries, consequently causing various environmental problems such as air pollution, water pollution, urban flooding, and urban heat islands. In order to monitor the urbanization process in these areas, remote sensing can be seen as a technology with great potential that has been proven in many previous studies in temperate regions. An accurate estimation of impervious surfaces in this region will have significant impacts in the long term to (1) provide better and more precise monitoring of impervious surface changes over time, (2) provide more precise information available for environmental models, which will foster the design of new models and improvement of existing ones, and (3) provide useful information to urban planners regarding locations where impervious surface fraction is too large, which areas are more prone to environmental problems, and so forth, leading to more eco-friendly urban planning.

One of the key instruments for mapping large areas of impervious surfaces is satellite remote sensing. However, accurate mapping of impervious surfaces remains a challenging task due to their diversity and the diversity of urban land covers, and thus impervious surfaces are often easily confused with other land cover types in terms of spectral signatures. For instance, bright impervious surfaces are often mixed with dry soils and sands, while dark impervious surfaces tend to be confused with shade and water. In order to reduce the spectral confusion, synthetic aperture radar (SAR) has been widely reported to provide complementary information. SAR works in all weather and time conditions, and thus is free from the influence of cloud occurrence. Moreover, existing research shows that SAR is very sensitive to ground surface roughness, shape, structure, and dielectric properties of illuminated ground targets, and thus can provide complementary information to optical data (Henderson and Xia 1997). Therefore, these characteristics and information should be of great benefit to separate different ground targets when their spectral signatures are similar in the visible and near-infrared wavelength range, as with bright impervious surfaces and dry soils/sands, or dark impervious surfaces and shade/water.

Unfortunately, in tropical and subtropical areas, remote sensing of impervious surfaces is much more complicated than in temperate urban areas due to several significant geographical factors:

1. *Complex meteorological environment.* Compared to mid- and high-altitude areas, the tropical and subtropical region, due to its special monsoon climate, presents a year-round cloudy and rainy climate feature with a very long rainy season and rich precipitation even in a short-term dry season. Natural disasters, such as typhoons, earthquakes, tsunamis, and floods, also occur frequently in this region because of its special meteorological environment. Therefore, the use of optical remote sensing would be greatly limited in tropical and subtropical areas. Microwave remote sensing, with its ability to penetrate clouds and operate in all weather conditions, has distinctive advantages in this region.

2. *Complex hydrological environment.* In tropical and subtropical areas, there are many developed and complex river systems and lakes in which the change of water flow and water surface area show obvious seasonal features, such as the big differences in variable source areas (VSAs) in rainy and dry seasons. This typical hydrological environment makes it difficult for remote sensing application. On remote sensing images, complex and discrete water surfaces can easily lead to the phenomenon of serious spectral confusion with other types of surface objects.

3. *Complex topographic environment.* Because of severe weathering and physical and chemical erosion, the topography in tropical and subtropical areas is complex, such as the distinctive Danxia landform,

karst landform, red weathering crusts, and krasnozem. All these complex topographies will result in problems of obvious shadows and buried structures on remote sensing images, which become a challenge in remote sensing monitoring. As well, complex topography and meteorological and hydrological environments are important factors of various natural disasters, such as karsts, landslides, and debris flows, which make it more difficult for remote sensing applications.

4. *Complex ecological environment.* Compared to a temperate zone, tropical and subtropical areas have the advantages of high ecological diversity and richer species resources. However, these diverse ecological features also make it difficult for the application of remote sensing. Low precision in remote sensing monitoring will occur because different species will very likely be mixed in with each other in remote sensing images. For example, surveys on ecological diversity using remote sensing techniques face many challenges due to the large number of species, such as the change detection of biodiversity in the mangrove forest that increasingly attracts public concern.

Moreover, in suburban-rural areas, where more natural and man-made features are mixed, ISE becomes more complicated due to the seasonal changes of vegetation. These seasonal effects have been identified previously (Weng et al. 2009; Wu and Yuan 2007), but an accurate ISE that would disprove any disruptive effects has yet to be designed.

This book therefore aims to (1) provide a systematic review of ISE methods using remote sensing (Chapters 1 and 2), by summarizing the environmental and socioeconomic impacts of impervious surfaces, the methods of ISE using remote sensing technology, and challenges of remote sensing in tropical and subtropical regions; (2) investigate the impact of climate zone and its seasonal effects on ISE (Chapter 4); (3) develop a framework for impervious surface estimation using optical and SAR image data (Chapters 3, 5, and 6); and (4) include an in-depth case study in rapidly urbanizing tropical and subtropical cities including Shenzhen, Mumbai, and Sao Paulo (Chapter 7).

1.2 Significance of Impervious Surface

As a key indicator of environmental studies, impervious surfaces have been considered extensively in terms of their environmental impacts (Arnold and Gibbons 1996). Built-up areas often have impervious surfaces, including pavement (roads, streets, highways, etc.) and rooftops, where precipitation water cannot infiltrate directly into the soil. Urban impervious surfaces can have a great impact on the urban solar energy balance (Weng et al. 2006), air quality, nonpoint source water pollution, storm runoff processes (Hurd and Civco 2004; Weng 2001), and so forth. In addition, impervious surfaces have also been identified as a key factor in many socioeconomic studies, such as the measurement of urban growth, the estimation of population distribution, and variation of housing prices (Wu and Murray 2003; Wu and Yuan 2007; Yang et al. 2003a; Yu and Wu 2004).

1.2.1 Environmental Significance

1.2.1.1 Hydrological Impacts

Impervious surfaces characterize all urbanized land and not only change the water cycles and the heat balance between the earth and the solar source, but also the living styles of people in different countries. As one of the results of these physical and social changes, impervious surfaces have had a significant impact on the environment, leading to a number of environmental issues all over the world. Regarding hydrological impacts, impervious surfaces change both the quantity and the quality of the watershed.

In a natural land surface, the general water cycle among hydrosphere, lithosphere, and atmosphere is as follows: water comes from the atmosphere with rainfall, infiltrates into the soils, and finally runs into rivers or underground water systems. During this process, surface runoff may be generated depending on the intensity of the rainfall, types of land covers, and soil types. Evaporation may also occur in different amounts depending on the land cover types and land surface temperature. However, in an urbanized surface that is covered by impervious materials, the water cycle changes dramatically. Rainfall water cannot infiltrate into the soil and is diverted into a drainage system. This water is actually surface runoff, and thus the runoff is greatly increased. All the rainwater from the drainage system is then transported into a river.

These changes to the water cycle would have significant impacts on the watershed in terms of water quantity (Jacobson 2011). The focus of these impacts is on flood occurrence during a precipitation event. First, floods come earlier in an urbanized area and the time of the flood peak also comes earlier (Espey et al. 1966). With the increase of surface runoff, the discharges of the floods are also increased (Espey et al. 1966) along with the drainage density and the flashiness of the storm flow (Graf 1977). Moreover, the flood duration is shortened as the flow runs much faster on impervious surfaces and in drainage tubes than on the soils (Seabum 1969). Meanwhile, during flooding, band erosion and the size of bed material is increased as the flow increases (Arnold et al. 1982). Several studies show that both the degree and spatial distribution of impervious surfaces have important influences on the water cycle (Booth and Jackson 1997; Brun and Band 2000; Sheeder et al. 2002). The water cycle only changes a little when the urbanization level is low, but the changes increase as urbanization increases. Booth and Jackson (1997) reported that if impervious surfaces are used as the indicator of urbanization, the changes begin to accumulate significantly after the percentage of impervious surfaces reaches 10%, while Brun and Band (2000) claimed that the threshold for impervious surfaces is about 20% in their study area. More recently, in another study conducted by Yang et al. (2010), they suggested 35% as a statistical threshold for impervious surface areas that would have a significant influence on the watershed. According to the literature, researchers agree that the appearance of impervious surfaces has significant impact on the watershed in terms of its quantity; however, studies are needed to investigate the exact impact of both the degree and the spatial distribution of impervious surfaces in urbanized regions, which is important for scholars of and decision makers within urban planning agencies and local governments.

In addition to water quantity, changes in the water cycle due to impervious surfaces also dramatically influence water quality, especially so-call nonpoint source (NPS) water pollution (Arnold and Gibbons 1996; Civico and Hurd 1997; Schueler 1994). Before urbanization, most pollutants that accumulated over land surfaces remained on the land because the intensity of surface runoff is low. Moreover, pollutants would be filtered by the soil during the infiltration process. However, as a consequence of urbanization, most pollutants accumulate on the roads, rooftops, and other impervious surfaces and are washed off by storms and transported by rainwater via drainage systems and finally into rivers. Studies have shown that various pollutants in an urbanized area are closely related to the impervious surfaces in the region (Bannerman et al. 1993; Schueler 1994). It has been reported that some important types of pollutants are highly related to impervious surfaces such as highways and roads. For instance, pathogens, nutrients, and toxic contaminants can be found on highways and roads, and these pollutants are harmful to the fish in the water body and can even cause harm to the animals and humans who drink the water directly or indirectly (Bannerman et al. 1993; Civico and Hurd 1997; Sleavin et al. 2000).

1.2.1.2 Urban Heat Islands

Impervious surfaces also change the heat balance between the land surface and the atmosphere. One well-known example is the urban heat island (UHI), which is an urbanized area that is much warmer than its surrounding rural areas (Weng et al. 2004). It is reported that the increase of temperature is correlated to the imperviousness of the urban area (Galli 1991). Impervious surfaces can influence the urban heat balance in several ways. First, most common impervious materials such as concrete and asphalt absorb more solar heat than natural land covers such as grassland and forest areas (Slonecker et al. 2001). This absorbed heat would then be released into the atmosphere and thus lead to an increase of the air temperature, which can be more than 10 degrees higher than the natural areas (Schueler 1994). Some governments are considering developing other types of impervious materials that are able to reduce the heat effects, such as the use of cool roofs consisting of materials that can highly reflect solar radiation from rooftops (Jo et al. 2010). Cool roofs are often done in cool colors such as white and light blue. The use of cool roofs can have a significant effect on reducing the UHI, as roofs account for more than 20% of the impervious surface areas in an urban city (Jo et al. 2010; Rose et al. 2003). Second, the occurrence of impervious surfaces reduces the coverage of vegetated areas, which actually plays an important role in balancing the heat in an urban area (Arnold and Gibbons 1996; Schueler 1994). Vegetation reduces the UHI effect by reflecting more solar radiation than impervious materials and consuming some solar energy with photosynthesis. This is why some cities are now building vegetated rooftops by planting various flowers or trees on roofs. Moreover, impervious surfaces also reduce the vegetation cover in stream sides (Hu 2011). These vegetation plants actually shade the stream and thus are able to reduce the temperature around the stream area. Klein (1979) reported the temperature in a nonshade stream can be up to 11 degrees higher than that in a shade stream in the Maryland area (Klein 1979). Much research has been conducted to investigate the quantitative relationship between impervious surface areas and the UHI (Lu and Weng 2006; Yuan and Bauer 2007).

1.2.2 Socioeconomic Significance

In addition to environmental impacts, impervious surfaces are also an important socioeconomic indicator. First, impervious surfaces have been considered as a better indicator compared to the number of urban population for urban sprawl evaluation (Torrens and Alberti 2000). Conventionally, population is the major parameter to assess the urbanization process. For instance, in China, the government uses the percentage of people who have registered as urban residences (according to the household registration system in China) in the total number of residences in an administrative area. That percentage is then used to characterize the percentage or degree of urbanization of the area. However, in a rapidly urbanized area (e.g., Guangzhou and Shenzhen in South China), a number of people who have registered with the government as having rural residences actually work and live in the city center. They may be local or external from other rural areas. The number of these people is actually very difficult to calculate; however, they should be counted as having urban residences because they work and live as urban citizens. Therefore, in this case, it becomes very difficult to quantize the urbanization degree of the studied area. Fortunately, impervious surfaces provide a good alternative for evaluating urbanization because from a physical point of view they account for the majority of changes in an urbanized area compared with how it was before becoming urbanized. Moreover, impervious surface distribution can be easily obtained either from a physical survey or by analyzing satellite images (Jat et al. 2008). Second, impervious surfaces are used to estimate population density. It is reported that impervious surfaces can be employed as supplementary data to estimate population density using a regression approach (Wang and Cardenas 2011). Wu and Murray (2005) used the cokriging approach to improve estimation by simultaneously accounting for the spatial autocorrelation of population density and impervious surface fraction (Wu and Murray 2005). Their research showed that population density and impervious surface fraction are coregionalized variables with low variance, and that impervious surfaces are better than other land use classes to estimate population density (Wu and Murray 2005).

1.3 Challenges of ISE

1.3.1 Land Cover Diversity and Spectral Confusion Issues

Land cover diversity is a direct challenge for land cover classification and ISE using remote sensing images. However, since land covers in urban areas and their changes are caused by both natural and human factors, land cover diversity is different from one place to another. It is important to figure out the challenges and impacts of land cover diversity in tropical and subtropical areas with regard to ISE. One of the most important problems produced by land cover diversity is the so-called spectral confusion, which refers to the similarity of spectral signatures among different land covers. Spectral confusion is also related to the difference of spectra within the different subtypes of one land cover type. Generally, there are some open issues related to the spectral confusion problem. First, bare soils or sands are often confused with bright impervious surfaces (e.g., cool roofs and new concrete roads), while shade and water are often confused with dark impervious surfaces (e.g., asphalt and old concrete roads). These confusions are caused by the similar spectral reflectance of different materials. Second, clouds and their shadows are considered a difficult issue to deal with in tropical and subtropical regions, where cloudy and rainy weather occurs throughout the entire year. Both of these problems lower the accuracy of the land use/land cover (LULC) classification in tropical and subtropical urban areas.

1.3.2 Scale Effects

The spatial resolution issue has long been considered very important in remote sensing applications (Jensen 2007; Jensen and Cowen 1999). During remote sensing of urban impervious surfaces, spatial resolution should be considered carefully, depending on the objectives of the study, for two reasons. First, coarse resolution data is more suitable for large-scale mapping of impervious surfaces due the availability of data and computational costs for data processing. For instance, for global mapping of impervious surfaces, coarse resolution data such as that from moderate-resolution imaging spectroradiometer (MODIS) and National Oceanic and Atmospheric Administration (NOAA) of about 1-km resolution is often selected, while Landsat data with 30 m resolution is becoming an emerging study area in recent years since the onset of global coverage data. However, very-high-resolution data (e.g., Quickbird and Ikonos) is not suitable for global remote sensing studies because global coverage data and continuous updates are not available for such a high resolution and the computational cost would be very high. Second, the methodology design and selection should be changed when using various resolutions of data for impervious surface mapping. There are different characteristics in the remote sensing data of different resolutions that require various methods to effectively extract accurate impervious surface information. In general, remote sensing data can be categorized into three types: coarse resolution, very high resolution, and high resolution. These three categories have previously been applied to estimate impervious surfaces.

Coarse resolution remote sensing data often refers to data with a spatial resolution that is lower than 100 m, such as advanced very high resolution radiometer (AVHRR) MODIS, and NOAA. Coarse resolution data is appropriate for regional and global mapping of impervious surfaces (Cracknell 1999; Gamba and Herold 2009; Lu et al. 2008; Quattrochi and Goodchild 1997; Schneider et al. 2010; Weng 2012). The selection of methods for processing coarse resolution data should depend on the mixed pixel phenomenon. Mixed pixels dominate coarse resolution images, and thus using per-pixel mapping of impervious surfaces often leads to overestimation or underestimation. Therefore, subpixel analysis methods such as spectral mixture analysis (SMA) are often adopted (Weng 2012).

Very-high-resolution remote sensing data often refers to data with a spatial resolution that is higher than 10 m, such as Quickbird, Ikonos, and Worldview. Local impervious surface mapping in a city may require very-high-resolution data since it can provide detailed LULC information that could be useful for urban planning and management. Much research has been done on remote sensing of impervious surfaces with very-high-resolution data (Cablk and Minor 2003; Goetz et al. 2003; Lu and Weng 2009; Lu et al. 2011; Wu 2009). Mixed pixels are significantly reduced in very-high-resolution data. However, there are challenging issues to address compared with coarse-resolution data. First, shadows from tall buildings and trees and topography are much more prevalent in very-high-resolution data than in coarse resolution data (Dare 2005). These shadows can influence the extraction of an impervious surface by reducing or removing the reflectance of solar radiation when using optical remote sensing data and the reflectance of active microwaves when using SAR remote sensing. Moreover, the spectral reflectance of shadows is often easily confused with dark impervious surfaces such as asphalt and old concrete roads and rooftops, and consequently, reduces the accuracy of impervious surface mapping. Second, spectral variations within the subtypes of land covers are increased in very-high-resolution images (Hsieh et al. 2001; Weng 2012). Therefore, the selection of methods should consider this characteristic. For instance, the frequently used maximum likelihood classifier (MLC) has been applied to classify remote sensing images. The application of MLC depends on the probability distribution of each class. However, in very-high-resolution images, the probability distribution of a land cover type may be changed as the spectral variation increases within the subtypes. In this case, the applicability of MLC should be reassessed before being applied to very-high-resolution images. In the processing of very-high-resolution remote sensing images, the combined use of spectral and spatial information is often recommended (Lu and Weng 2007).

Remote sensing images with a spatial resolution between 100 and 10 m can be treated as high-resolution images, such as Landsat and Système Pour l'Observation de la Terre (SPOT). High-resolution images should be the most frequently used data due to the great range of applications and availability of datasets. For instance, Landsat data has been applied in many global remote sensing studies such as global impervious surface mapping and global LULC mapping. However, high-resolution images have very complicated image characteristics, including mixed pixels, shadows from tall buildings, and topography and spectral variation within one land cover type and subtypes. Therefore, the selection of methods depends on a number of factors, such as the landscape of the study area, the seasons of the selected data, and the application objectives of the study.

1.3.3 Influence from Climatology and Phenology

There are several types of climate in tropical and subtropical areas, such as tropical rainforest climate, tropical wet-dry climate, and humid subtropical climate.

1.3.3.1 Humid Subtropical Climate

A humid subtropical climate is known as Cfa or Cwa in the Koppen climate classification of world climate (Peel et al. 2007). This climate is characterized by hot, humid, and generally mild to cool winters (Peel et al. 2007). The Cfa or Cwa regions are generally located on the east coast of continents between 20° and 40° North and South latitude. Peel et al. (2007) demonstrate the spatial distribution of humid subtropical climates all over the world. Therefore, a humid subtropical climate can be found in the southeastern parts of the United States, South America, and South Africa, and the eastern parts of Australia and Asia (e.g., Northern India, China, and Japan).

A humid subtropical climate is characterized by its unique temperature and precipitation features. Generally, the temperature is between 21°C and 26°C in summer and above 0°C in winter. Moreover, the variance of temperature in one day is very small. Thus, there is warm winter and very little temperature difference between summer and winter. On the other hand, a humid subtropical climate experiences high precipitation and is wet and humid throughout the year. There are dry and wet seasons. In general, the wet season begins from May to August when there is a lot of rainfall. The dry season begins from September to January when the number of cloudy and rainy days is reduced. From February to April, the weather is warm and humid.

1.3.3.2 Humid Subtropical Phenology

Generally, a humid subtropical phenology is determined by the humid subtropical climate through the characteristics of temperature and precipitation. The phenology is widely recognized to strongly depend on the climate (Hudson et al. 2009; Inouye 2008; Jones and Davis 2000). In terms of humid subtropical areas, previous studies suggest that there is strong seasonality of the plants in subtropical forest ecosystems (Kikim and Yadava 2001). Generally, the natural vegetation in this zone is a subtropical evergreen forest, which occurs in two forms: broadleaf and needle-leaf. There is no obvious difference between spring (leaf-on) and autumn (leaf-fall) as in the temperate regions (Shukla and Ramakrishnan 1982). However, there are some differences in the forest in different seasons due to the growing season when leaves are flushing and some tree species are flowering. These differences can be reflected in the concentration of chlorophyll in the leaves, which can be observed by remotely sensed data. As a result of this seasonal variation of plants, vegetation cover changes seasonally over a large scale on the land surface, which can also be seen from satellite images.

1.3.3.3 Seasonal Effects from Climatology and Phenology

The seasonal changes of the climatology and phenology can result in seasonal changes of land covers. These seasonal effects include two aspects: (1) the difference in precipitation in dry seasons and wet seasons can produce differences of water surface area on the land surface, and (2) the seasonal changes of plants will change the vegetation coverage in the areas of hills, mountains, and greening zones in urban areas. Water body and vegetation are two important land cover types in urban remote sensing studies, and the distribution of these two land cover types can lead to different patterns of spectral confusions in a given urban study area.

The land cover changes produced by the climatology and phenology are the so-called seasonal effects detailed in Chapter 4. As pointed out in previous research, ISE from satellite images vary in different seasons due to the seasonal changes of vegetation. Impervious surfaces were reported to be overestimated in wintertime, when tree canopies were at their minimum (Weng et al. 2009). Moreover, different ISE methods may be sensitive to this seasonal change in different ways. For instance, linear spectral mixture analysis is more sensitive to seasons, while the regression tree model is less sensitive to this change (Wu and Yuan 2007). Nevertheless, the work done by Wu and Yuan (2007) and Weng et al. (2009) was conducted at the mid-latitude region where the plant phenology undergoes dramatic changes in different seasons (Weng et al. 2009; Wu and Yuan 2007). Thus, it is still not clear whether the seasonal effects observed previously suit the situation in subtropical humid areas such as the Pearl River Delta (PRD). Chapter 4 aims to address this question and assess the seasonal effects of ISE in the PRD region.

Seasonal effects have been identified as an important factor for ISE with regard to remote sensing data, but the seasonal effects in tropical and sub-troprical regions remain unclear. Seasonal changes of land covers from remote sensing images can be caused by various factors, such as plant phenology and seasonal precipitation changes. In those cases, the seasonal land cover changes would influence the accuracy of classification due to the differences of spectral confusions between land covers with a similar reflectance. Since plant phenology and precipitation are unique in tropical and subtroprical regions compared with other study areas in the literature, the seasonal effects should be reexamined to select the best season for ISE. The hypothesis is that plant phenology and climatology in tropical and subtroprical regions will determine the best season for ISE from satellite images, which would be different than in other regions of the world, and that winter may be the best season for imaging because it is dry with less clouds and precipitation.

1.3.4 Multisensor Fusion

Most previous studies were focused on using a single data source, mainly optical remote sensing, to estimate impervious surfaces. However, because of the rapid development of advanced satellites and the availability of multiple sources of satellite data, there is an increasing demand for multisensor fusion to obtain more accurate ISE at local, regional, and global scales. Among these multisensor applications, light detection and ranging (lidar) and SAR data are two commonly used additional data sources. Moreover, since there is still no spaceborne lidar data, SAR is actually the most frequently used data source to be incorporated with optical images for ISE.

SAR has been widely recognized as an important data source that is able to compensate optical remote sensing images in urban remote sensing studies. However, comprehensive assessment of SAR data for improving the estimation of urban impervious surfaces is still insufficient. In addition, the existing literature focuses more on the advantages of SAR data and less on the disadvantages. In order to effectively combine the two data sources, it is important to understand both the advantages and disadvantages of SAR data. The hypothesis about this issue is that SAR provides complementary information with the advantage of being sensitive to the urban surface roughness; nevertheless, the disadvantages of SAR data regarding ISE should also be recognized and taken into consideration.

Technically combining optical and SAR images for accurate ISE is one of the key issues in this book. Guided by the understanding of the two data sources and the challenges of ISE, the design of the methodology (including feature extractions from two data sources, selection of fusion levels, and classification models) is to be proposed, implemented and validated in this research. The technical hypothesis is that feature extraction for both the optical and SAR images are important for the fusion, but the two data sources should be treated in different ways during the fusion procedure.

1.4 Objectives and Significance

The overarching goals of this book are to review the state of the art of ISE using remote sensing and to develop a generic methodology for accurate ISE from multisatellite images in tropical and subtropical areas. The specific objectives of the book are therefore to

1. Summarize the environmental and socioeconomic impacts of impervious surfaces, the state-of-the-art methods of ISE using remote sensing technology

2. Investigate and assess the impact of climate zone on ISE in tropical and subtropical areas

3. Investigate and assess the diversity of land covers in a rapid urbanized area of tropical and subtropical regions

4. Explore the additional use and potentials of radar images for improving the accuracy of ISE

5. Synergistically use both the optical and radar images for accurate estimation of urban impervious surfaces in tropical and subtropical areas

Impervious surfaces have increased dramatically over the past decades as a result of rapid urbanization in tropical and subtropical countries. It has been widely recognized that impervious surfaces serve as a key environmental indicator since they affect the water cycle, water pollutants, and the energy balance, and thus, relate closely to the UHI effect. It has also been reported that impervious surfaces serve an important role in urban socioeconomic studies such as detailed population distribution.

Given their importance, many methods have been developed to assess impervious surfaces using satellite images, but most were tested in temperate, continental areas. Few considered the cases in tropical and subtropical areas where there is significant cloud occurrence and different phenology. Due to these differences, difficulties are encountered when using optical remote sensing data—the main data source used in previous studies—in tropical and subtropical areas. Seasonal changes in vegetation, recognized as a key factor in ISE, create further challenges. Existing approaches need to be evaluated while additional data sources should be considered in order to establish an advanced assessment of impervious surfaces in tropical and subtropical urban areas.

The outcome of the research detailed in this book will provide evidence of the seasonal effects on impervious surface assessment due to phenological changes, evaluate the potential of SAR data for improving impervious surface assessment, and design a comprehensive framework to estimate impervious surfaces using optical and SAR images. As well, the methodology and conclusions of this research will serve as a general and useful reference for urban remote sensing studies in tropical and subtropical countries.

1.5 Organization of This Book

The organization of this book is briefly illustrated in Figure 1.1. This chapter has introduced the background of existing research, the research questions, issues, and related hypotheses, and the objectives and significance of the research. Chapter 2 reviews the literature on the significance of ISE and the phenological and climatic characteristics of tropical and subtropical regions, as well as on previous research of ISE. Chapter 3 describes the study materials in Chapters 4 through 7, including study areas, test sites, and datasets such as satellite data, digital orthophoto, and *in situ* data. A methodological framework is also presented in Chapter 3, including methods for investigating the seasonal effects of ISE, feature extraction methods, the fusion between optical and SAR data, and the validation methods of the results. The results are presented and discussed in Chapters 4 through 7, presenting and discussing the results of feature extraction based on shape-adaptive neighborhood (SAN), the seasonal effects of ISE, urban land cover diversity, and the combined use of optical and SAR data, respectively. Chapter 8 summarizes the main findings and conclusions, as well as the limitations and future research.

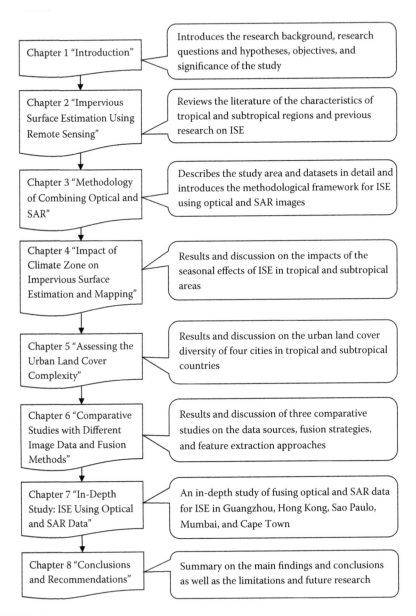

FIGURE 1.1
Organization of this book.

2

Impervious Surface Estimation Using Remote Sensing

2.1 Overview of the Methodology

Urban impervious surfaces can have a great impact on the urban solar energy balance, air quality, nonpoint source water pollution, storm runoff process, and so forth. In addition, impervious surfaces have also been identified as a key factor of many socioeconomic issues such as population distribution and house price. Thus, accurate mapping of impervious surfaces becomes significant not only for the environmental monitoring but also for various socioeconomic studies. In some developed countries, much work has been done to map impervious surfaces. For instance, in the United States, the impervious surfaces layer was developed using Nighttime Lights Time Series (2000–2001) and LandScan 2004 population count, at a coarse resolution, along with United States Geological Survey (USGS) 30 m resolution ISE of the United States for calibration (Elvidge et al. 2007; NOAA 2010). In Europe, there is a series of seamless European mosaics for impervious surfaces (EEA 2006), including an updated raster dataset containing the degree of imperviousness ranging from 0%–100% in aggregated spatial resolution (100 × 100 m).

Dramatic urbanization has taken place in many regions creating a number of metropolises in the world. The PRD region is one such area. The region is located on the Pearl River Estuary (PRE), known as the third largest metropolitan area in China, experiencing tremendously fast development during the past 30 years. The region has rapidly become urbanized, with a population of over 19 million in an area of over 21,000 km^2 (Fan et al. 2008). One of the major features of such rapid urbanization is the increase in the spread of impervious surfaces.

Hong Kong and other cities in PRD are now covered by large areas of impervious surfaces, including rooftops, transportation networks, and parking areas. Kowloon and the north of Hong Kong Island is one of the most heavily urbanized areas in the world. Previous research indicates that the percentage of impervious surfaces in the commercial business districts and residential areas on both sides of Victoria Harbour has exceeded 70% (Jiang et al. 2009).

2.2 ISE Using Single Data Source

2.2.1 Spectral Mixture Analysis

Spectral mixture analysis (SMA) is a conventional approach for impervious surface estimation at the subpixel level. SMA is based on the biophysical composition model called vegetation-impervious surface-soil (VIS). By excluding the water surface, which is relatively easier to identify, the VIS model treats the urban environment as a biophysical composition of three components: vegetation, impervious surface, and bare soil (Ridd 1995; Wu and Murray 2003). Therefore, a pixel in a remote sensing image can be treated as a combination of these three components. With a spectral mixture model, the percentage of each component can be calculated. The spectral mixture model describes how the three components are combined or mixed together to form a final pixel. There are both linear and nonlinear spectral mixture models. The selection of models should depend on the complexity of land covers. If each photon interacts with single land cover type, a linear mixture model should be applied, but if each photon interacts with multiple land cover types, a nonlinear model should be applied (Wu and Murray 2003). However, a composition model can be approximately treated as a linear model in an urban environment where the nonlinear effect can be neglected (Phinn et al. 2002; Rashed et al. 2001; Small 2001, 2002). Therefore, linear spectral mixture models are often applied in urban impervious surface estimation. SMA using a linear spectral mixture model is also called linear SMA (LSMA). Wu and Murray (2003) divided the impervious surface into two subtypes: low albedo impervious surface and high albedo impervious surface (Wu and Murray 2003). Therefore, the VIS model was extended to include four components: vegetation, low albedo impervious surface, high albedo impervious surface, and bare soil. By solving the SMA or LSMA models, fractions of each component can be obtained, and then, the impervious surface percentage can be calculated with the fractions of subtypes of impervious surfaces and the low albedo and high albedo fractions. A typical working flow of impervious surface estimation using LSMA is described below.

2.2.1.1 Endmember Selection

Theoretically, optimal endmembers should be selected by laboratory measurement of each endmember's spectra. When applying them to remote sensing studies, remote sensing images should be corrected to remove the atmospheric effects, and then the linear mixture analysis can be applied using the laboratory endmember data and corrected remote sensing data. However, in real applications, lab-based measurement of endmembers is probably unavailable, and thus endmembers should be derived from remote sensing images. One popular approach of selecting endmembers from remote sensing images is to use the visualizing spectral scatter plots of different image bands or the components of the image transformation (e.g., principal component [PC] and maximum noise fraction [MNF]). To show how this approach is applied for endmember selection, an example proposed by Wu and Murray (2003) is given here. In Wu and Murray's research, the MNF transformation was employed to minimize the noises in one band and in band-to-band correlation. The three steps of MNF are (Wu and Murray 2003): (1) applying the PC transformation to diagonalize the noise covariance matrix, (2) converting the noise covariant matrix to an identity matrix, and (3) performing a second PC transformation. The first two components were identified to have significant spatial variations between different land covers, and the third component showed a significant feature for bare soil. Then, the scatter plots of these MNF components can be visualized to guide the selection of endmembers. In their research, the first three components of MNF transformation were used, and every pair of components was plotted indicating the locations of the distributions of different land cover types. Then four endmembers (high albedo, low albedo, vegetation, and soil) were identified and selected by comparing this feature space and their association in the original reflectance image. After selecting the endmembers, the reflectance of each endmember can be calculated.

2.2.1.2 Linear Spectral Mixture Model and Fraction Images

After selecting the endmembers, LSMA can be applied using the linear spectral mixture model. LSMA is a physically based image processing approach that describes the linear composition of spectra within a pixel in an image using the spectra of endmembers. Detailed descriptions of the linear spectral mixture model and its principles can be found in the literature (Adams et al. 1995; Roberts et al. 1998; Weng et al. 2008; Wu and Murray 2003). The basic mathematical model can be expressed as

$$R_b = \sum_{i=1}^{N} f_i R_{i,b} + e_b \tag{2.1}$$

where R_b is the reflectance of a given pixel in Band b, $R_{i,b}$ is the reflectance of endmember i in Band b in the case of a pure pixel, f_i is the fraction of endmember i in the pixel, N is the number of endmembers, and e_b is the unmodeled residual. Moreover, there are two constraints, $\sum_{i=1}^{N} f_i = 1$ and $f_i \geq 0$. This is also known as the fully constrained linear mixture model. In a real application after endmember selection, R_b and $R_{i,b}$ are known, and the objective is to solve out f_i that is the fraction of each endmember. To solve the model, besides the abovementioned two constraints, there is another constraint to minimize the overall unmodeled residual in all bands, which can be expressed as root-mean-square (RMS) error:

$$RMS = \left(\sum_{b=1}^{M} e_b^2 / M \right)^{1/2} \tag{2.2}$$

where M is the number of bands. However, it is difficult to perfectly solve the fully constrained linear mixture model subjected to the constraints. Therefore, in order to find a better solution, three conditions should be satisfied (Weng et al. 2008): (1) the selected endmembers should be independent, (2) the number of endmembers should not be greater than the number of bands, and (3) the reflectance of bands should not be highly correlated.

After solving the linear spectral mixture model, the fraction of each endmember in each pixel can be obtained. By showing the fraction of all the pixels, a fraction image can be generated for each endmember, and the RMS image shows how well the full constrained spectral mixture model is solved. In Wu and Murray's research, the mean RMS is 0.0057, indicating a generally good fit (less than 0.02). RMS in residential, vegetation, soil, and water cover types is rather low. However, some high albedo materials, such as high-reflectance roofs, clouds, and sand, are not so good with higher RMS values.

2.2.1.3 Impervious Surface Estimation

The modeling of an impervious surface is not straightforward due to its complexity in spectral reflectance. However, Wu and Murray (2003) found that impervious surfaces can be approximately modeled by adding the low albedo and high albedo fraction image. To test this conclusion, the following experiment has been done by both Wu and Murray (2003) and Weng et al. (2009). In a central business district (CBD) region where all the pixels are supposed to be impervious surfaces, the reflectance of impervious surfaces can be expressed as

$$R_{imp,b} = f_{low} R_{low,b} + f_{high} R_{high,b} + e_b \tag{2.3}$$

where $R_{imp,b}$ is the reflectance of the impervious surface of a given pixel in Band b, $R_{low,b}$ is the reflectance of the endmember low albedo and f_{low} is its corresponding fraction in the pixel, $R_{high,b}$ is the reflectance of the endmember high albedo, f_{high} is its corresponding fraction in the pixel, and e_b is the unmodeled residual. This model is constrained to $f_{low} + f_{high} = 1$ and $0 < f_{low}, f_{high} < 1$. In their experiments, both results show good fitting of this two-endmember spectral mixture model with the impervious surface located near the line connecting low albedo and high albedo endmembers.

Therefore, the fraction of an impervious surface can be simply considered as the sum of the fractions of low albedo and high albedo. By doing so, a fraction image of an impervious surface was generated in Wu and Murray's study.

Lastly, in order to successfully model the impervious surface with LSMA, nonimpervious surfaces with low albedo and high albedo should be considered and removed before applying the LSMA. Generally, a low reflectance nonimpervious surface includes water and shaded areas, while a high reflectance nonimpervious surface includes cloud and sands. Only with all these pixels removed can the LSMA be used to model the impervious surface successfully (Wu and Murray 2003; Weng et al. 2008).

2.2.2 Normalized Impervious Surface Indices

Normalized impervious surface indices (NISIs) have been developed to enhance the impervious surface in a remote sensing image, working similarly to the well-known normalized difference vegetation index (NDVI). So far, there are two normalized indices that can be used to characterize impervious surfaces, the normalized difference impervious surface index (NDISI) developed by (Xu 2010) and the biophysical composition index (BCI) developed by (Deng and Wu 2012). The basic ideas of these NISIs are described next.

2.2.2.1 NDISI

The general idea of NDISI is to enhance the difference of spectral reflectance of impervious surfaces and nonimpervious surfaces in the visible, near-infrared (NIR), mid-infrared (MIR), and thermal infrared (TIR) ranges. The basic idea of NDISI relies on the following observations (Xu 2010): (1) impervious materials (e.g., concrete and asphalt) have a stronger capability of emitting heat reflected in TIR and lower reflectance in NIR, (2) soil, sand, and water generally have higher reflectance in visible bands, and (3) soil and sand have stronger reflectance in the MIR band. Based on these observations, the NDISI can be calculated as

$$NDISI = \frac{TIR - [(VIS_1 + NIR + MIR_1)/3]}{TIR + [(VIS_1 + NIR + MIR_1)/3]} \qquad (2.4)$$

where TIR is the thermal band, VIS_1 is the reflectance of visible band, and NIR and MIR_1 are the reflectance of the NIR and MIR bands. However, Xu's experiment showed that water can have a lower reflectance than an impervious surface in visible bands in clear waters, causing some confusion in the NDISI values between water and impervious surface (Xu 2010). In order to address this problem, Xu found that the use of water index (WI) instead of visible band was able to reduce this confusion, as WI can enlarge the contrast between water and an impervious surface (Xu 2010). Therefore, NDISI can be calculated as

$$NDISI = \frac{TIR - [(WI + NIR + MIR_1)/3]}{TIR + [(WI + NIR + MIR_1)/3]} \qquad (2.5)$$

where *WI* can use any water index such as the NDWI proposed by McFeeters (1996) and the modification of normalized difference water index (MNDWI) developed by Xu (2006).

$$MNDWI = \frac{(Green - MIR_1)}{(Green + MIR_1)} \qquad (2.6)$$

$$NDWI = \frac{(Green - NIR)}{(Green + NIR)} \qquad (2.7)$$

where *Green* is the reflectance in the green band.

Finally, the NDISI value is a real number between –1 and 1, with higher values indicating impervious surfaces and lower values indicating nonimpervious surfaces. Therefore, NDISI means neither the percentage nor the area of an impervious surface. In order to estimate the percentage or area of an impervious surface, more processing should be employed using NDISI. For instance, Xu (2010) used multivariate statistical analysis to estimate the subpixel percentage of impervious surfaces, with the help of training samples from higher-resolution images. Lastly, it should be noted that the TIR band is required to calculate NDISI. Therefore, NDISI cannot be calculated for some remote sensors without TIR data such as SPOT, Ikonos, and Quickbird.

2.2.2.2 BCI

BCI was recently proposed to characterize urban biophysical composition and identify impervious surfaces based on the normalized tasseled cap (TC) transformation (Deng and Wu 2012). According to the BCI principles, an impervious surface falls into the range of positive and higher BCI values, while vegetation and soil have negative BCI values. The original idea of BCI follows the VIS triangle model proposed by (Ridd 1995). In order to transform the BCI to be positive for impervious surfaces and negative for nonimpervious surfaces, Deng and Wu (2012) employed and examined the components of TC transformation. Two important observations were found in their experiment: (1) TC1 (first TC component) and TC3 (third TC component) showed a strong negative linear relation by an elongated strip, (2) a higher value of TC3 does not always corresponds to higher water concentration. Therefore, the first three TC components were explained to correspond to three typical urban biophysical compositions: TC1 as "high albedo," TC2 as "vegetation," and TC3 as "low albedo." Moreover, if the TC components are normalized to be between 0 and 1, these relationships are clearer; that is, TC1 is highly related to bright impervious surface, TC2 is highly related to vegetation, and TC3 is highly related to a dark impervious surface. Based on these observations, the BCI can be calculated with the following equation:

$$BCI = \frac{(H+L)/2 - V}{(H+L)/2 + V} \qquad (2.8)$$

where H is the normalized TC1 as high albedo, L is the normalized TC3 as low albedo, and V is the normalized TC2 as vegetation. The normalized TC components can be computed as follows:

$$H = \frac{TC1 - TC1_{min}}{TC1_{max} - TC1_{min}}$$

$$V = \frac{TC2 - TC2_{min}}{TC2_{max} - TC2_{min}} \qquad (2.9)$$

$$L = \frac{TC3 - TC3_{min}}{TC3_{max} - TC3_{min}}$$

where TCi ($i = 1, 2, 3$) are the first three components and TCi_{min} and TCi_{max} are the minimum and maximum values of the ith TC components.

Compared with NDISI, BCI does not depend on TIR bands and it can be applied to any remote sensing images theoretically. However, note that BCI cannot represent the percentage or area of an impervious surface; additional methods should be used to quantify it. For instance, Deng and Wu (2013) conducted other research on impervious surface estimation using BCI at the subpixel level. In their research, BCI was used to help select endmembers in an automatic way in order to improve the accuracy of impervious surface mapping using LSMA (Deng and Wu 2013).

2.2.3 Artificial Neural Network (Multilayer Perceptron and Self-Organizing Map)

2.2.3.1 Multilayer Perceptron

One of the most widely used artificial neural networks (ANNs) is the multilayer perceptron (MLP) feedforward network (Kavzoglu and Mather 2003), which is structured with three types of layers: input, hidden, and output. For the application of satellite imagery, the input layer corresponds to the bands of multispectral images, and the output layer represents different kinds of land use, land cover, or various material classes (Weng and Hu 2008). The different procedures about classifying are conducted within the hidden layer(s), depending on both the number of hidden layers and the number of nodes in each hidden layer. In this study, the input layer corresponded to the six bands (30 m × 30 m) for the enhanced thematic mapper plus (ETM+) image. For the advanced synthetic aperture rader (ASAR) data, speckle filtering and texture analysis with the gray level co-occurrence matrix (GLCM) method in Section 3.5.3 was first applied, and then the input layer corresponded to the texture images of the ASAR data. A typical structure of the three-layer MLP is shown in Figure 2.1.

Except for the input layer and the output layer, the nodes in the hidden layer will conduct the classification procedures and each node has a similar structure shown as in Figure 2.2.

The function *f* is called the activation function, which can be illustrated as the following formula:

$$O_j = f\left(\sum_j W_{ij} I_i\right) \tag{2.10}$$

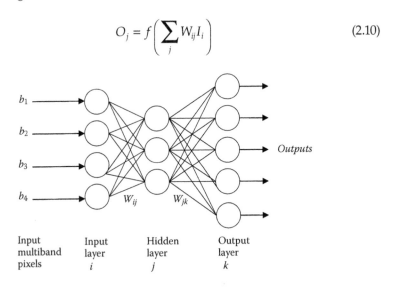

FIGURE 2.1
Structure of MLP.

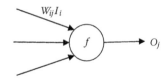

FIGURE 2.2
A node in the hidden layers.

where W_{ij} denotes the weighting of the previous output (I_i) from an input node, and O_j is the output of this node.

The first consideration is the number of hidden layers. There may be one or two hidden layers for the application of MLP in remote sensing classification (Weng and Hu 2008). However, a single hidden layer should be sufficient for most problems, especially for classification tasks (Cybenko 1989; Lippmann 1987). This is considered to be appropriate for most multispectral satellite images since there are ordinarily only several spectral bands. Staufer and Fischer (1997) pointed out that if the optimum number of hidden nodes on a single layer is larger than 20, another hidden layer may be needed and a portion of the nodes can be moved to the other hidden layer (Staufer and Fischer 1997).

Another key factor is the number of nodes. Various strategies have been discussed to determine of the number of nodes for each hidden layer, but few of them are widely accepted (Richards and Jia 2006). One approach taking into account both the number of input nodes and output nodes seems to be more suitable for remote sensing classification (Eastman 2003; Weng and Hu 2008). From their research, the number of nodes in the hidden layer can be estimated as

$$N_h = INT\sqrt{N_i \times N_o} \tag{2.11}$$

where N_h denotes the number of nodes in the hidden layer, N_i is the number of nodes in the input layer, and N_o is the number of nodes in the output layer.

After the structure is determined, a learning algorithm should be designed and related parameters should be set, so that the MLP can learn the prior knowledge from the training data set and obtain a better ability for classifying other datasets. Thus, the learning method is also a key factor for the success of the MLP. The backpropagating (BP) learning algorithm is a widely used approach that employs the generalized delta rule (Richards and Jia 2006). After the learning/training process, all the knowledge about the classes should be contained in the weights of all the nodes in the MLP model. Finally, in order to optimize the learning process and the classifying procedure, different parameters should be given to the MLP, including the learning rate, the momentum factor, and the threshold for the accuracy level (Richards and Jia 2006; Weng and Hu 2008).

2.2.3.2 SOM

SOM is another type of ANN and is also a data clustering technique. SOM has been successfully applied to estimation of impervious surfaces from remote sensing images (Hu and Weng 2009). Unlike MLP, SOM has only two layers, the input layer and the output layer. The input layer corresponds to the input features vector such as the image bands, while the output layer consists of a 2-D array, which is often a square. The output layer is a competitive layer where each neuron corresponds to a class. Between the input layer and the output layer, each neuron in the output layer is connected to all the neurons in the input layer, and the connection weight between each pair of neurons falls into [0, 1].

During the learning stage, the construction of a SOM includes four main steps (Hu and Weng 2009). First, a coarse tuning process is conducted to determine the weights of all the connections between the neurons in the input layer and output layer. Second, a labeling procedure is carried out to assign one class label to each neuron in the output layer. The majority voting approach using training dataset is often used to determine the class label of each neuron. Third, a fine-tuning process is conducted to improve the connection weights in order to increase the discriminability of the decision boundaries and the learning vector quantization (LVQ) approach is often adopted. Finally, the trained SOM is built and can be applied to predict the class label of unknown data.

The construction of an effective SOM lies in several important factors. First, the map size of SOM in the output layer should be determined. In the study of Hu and Weng (2009), a size of 4 × 4 was used by testing a number of different map sizes. Other parameters should be set manually including initial neighborhood radius, minimum learning rate, and maximum learning rate. Second, there are different methods during the tuning processes to trigger the competitive layer neurons using the connection weights and the input features. Two commonly used methods, including SOM commitment (SOM-C) and SOM typicality (SOM-T), were tested for impervious surfaces estimation previously (Hu and Weng 2009).

2.2.4 Support Vector Machine

Unlike conventional empirical risk minimization (ERM) methods such as the ANN, the support vector machine (SVM) originally comes from the structural risk minimization (SRM) principle proposed by (Vapnik 1995). The basic idea of the SVM is to map multidimensional data into a higher dimensional space, in which there is a hyperplane that can be used to linearly separate the original data, thereby maximizing the margin between different classes (Vapnik 1998). The theory of SVM has been described extensively in previous reviews (Hsu et al. 2007; Vapnik 1998; Weston and Watkins 1999), and the SVM approach employed in this chapter was described by Hsu et al. (2007). In particular, SVM can work without a kernel function if the data is linearly separable. Otherwise, a kernel function is needed to map the data into a higher dimensional space where the data is linearly separable. The success of classification using SVM was recognized to depend on the parameter of the penalty (C) and the Gamma (G) in the kernel function.

First, the training data can be labeled as $\{x_i, y_i\}$, $i = 1, ..., l$, $yi \in \{-1, 1\}$, $xi \in R^d$, where x_i is a vector of the data, like the spectral values for each pixel, and y_i

is its class type for this pixel. For the linearly separable case, a hyperplane should exist and satisfy $\mathbf{w} \cdot \mathbf{x} + b = 0$, where $\mathbf{w}$ is normal to the hyperplane. Moreover, $|b|/\|\mathbf{w}\|$ is the perpendicular distance from the hyperplane to the origin, and $\|\mathbf{w}\|$ is the Euclidean norm of $\mathbf{w}$. Therefore, all the data in this hyperplane should satisfy the following constraints:

$$\mathbf{xi} \cdot \mathbf{w} + b \geq +1 \quad \text{for} \quad y_i = +1$$
$$\mathbf{xi} \cdot \mathbf{w} + b \leq -1 \quad \text{for} \quad y_i = -1$$

(2.12)

Then these two constraints can be combined into a single one:

$$y_i (\mathbf{xi} \cdot \mathbf{w} + b) - 1 \geq 0$$

(2.13)

The goal of SVM is to find such a hyperplane that creates the maximum margin between two classes of data, and this hyperplane is called the optimal separating hyperplane (OSH). The procedure of finding the OSH in the training process can be represented as the following quadratic optimization problem:

$$\min \frac{1}{2} w^2 + C \left(\sum_{i=1}^{l} \xi_i \right)$$

$$\textit{s. t.}$$

$$\begin{cases} y_i [WX_i + b] \geq 1 - \xi_i \\ \xi_i \geq 0, i = 1, 2, \ldots, l \end{cases}$$

(2.14)

To solve this optimization problem, it can be converted into an equivalent Lagrange dual problem (LDP), with the following form:

$$W(\alpha) = \sum_{i=1}^{l} \alpha_i - \sum_{i,j=1}^{l} \alpha_i \alpha_j y_i y_j K(x_i, x_j)$$

(2.15)

with the constraints: $\sum_{i=1}^{l} \alpha_i y_i = 0, 0 \leq \alpha_i \leq C$, and $i = 1, 2, \ldots, l$.

After solving the problem, a final decision about other data can be made with the following function:

$$f(x) = sign \left[\sum_{i=1}^{n} \alpha_i y_i K(x_i, x_j) + b \right]$$

(2.16)

In the above equations, α_i represents the Lagrange multipliers, C is the penalty for misclassification, and n is the number of support vectors that determine the OSH. $K(x_i,x_j)$ is the kernel function that can map the data into a higher dimension. There are three classical kernel functions often used: the polynomial kernel function, the radial basis kernel function, and the sigmoid kernel function. The success of classification using SVM has been recognized to depend on the parameter of the penalty (C) and the Gamma (G) in the kernel function.

The SVM described above is the binary SVM and is only suitable for separating two classes. Thus, a multiclass version of the SVM is needed for the classification task in our case. Actually, there have been different approaches converting the binary SVM into a multiple-class version, and among them the one-against-rest method has been widely used to solve this type of problem (Weston and Watkins 1999). The basic idea is to use multiple binary SVMs, and each would consider a certain class as the first class while treating other classes as the second class. We can then get one hyperplane from each binary SVM, and by combining all these hyperplanes, all classes can be separated in the hyperspace.

In this book, the same optical and SAR data used in the MLP was used as an input for the SVM in order to compare the effectiveness of these two methods. For the ETM+ image, the six bands (30 m × 30 m) were input into the SVM. For the ASAR data, speckle filtering and texture analysis with the GLCM method in Section 3.1 was also applied to obtain the texture features of the ASAR data. These texture images were then used as input for the SVM.

2.2.5 Classification and Regression Tree

The basic idea of a classification and regression tree (CART) is to grow a binary tree by a recursive partitioning process starting from a tree node (Breiman et al. 1984). CART is built by growing two children nodes from one parent node recursively. The Gini index is often used to measure the impurity of a node when building up the node (Breiman et al. 1984). CART can be applied in a classification task or regression task depending on categorical attributes or numeric attributes. CART has been applied to remote sensing image classification and was proven to obtained good accuracy with good predictability (Huang et al. 2002). The application of CART in impervious surfaces was first conducted by Yang et al. (2003a) in a regression task to estimate the percentage of impervious surfaces. Their research concluded that CART was able to obtain consistent and acceptable accuracy in three different study areas (Yang et al. 2003a). Moreover, it was found that CART could finish the task in only a limited computing time, indicating its great potential for large-scale mapping of impervious surfaces (Yang et al. 2003a).

2.2.6 Object-Oriented Analysis

Object-oriented analysis, also known as object-based analysis, was proposed for very-high-resolution remote sensing image classification (Benz et al. 2004). Its successful applications compared with pixel-based methods lies in the basic idea of analyzing and classifying the remote sensing images based on the object level, which produces good classification results without the noticeable noise phenomena that are common in those results from pixel-based classification. There are typically three steps in an object-oriented analysis. First, it begins with the segmentation of images by dividing the images into a set of separated regions or objects, which are composed of a set of pixels. Second, feature extraction is applied to all the objects to extract various features such as spectral features (e.g., reflectance mean and NDVI) and spatial features (e.g., texture features, shape features, and topographical features). A number of methods are available for extracting both spectral and spatial features from the objects. Third, a classification procedure is conducted on the objects to group them into different classes based on the features extracted during the second step. Commonly used classification methods include membership function approach and the nearest neighbor classifier. The applications of object-oriented analysis on urban land cover classification and impervious surface estimation have been conducted in some studies (Hu and Weng 2009; Lee and Warner 2006; Myint et al. 2008, 2011). It was proven that object-oriented analysis was superior to pixel-based classification approaches for high-resolution images.

The successful application of object-oriented classification depends on many factors, such as the settings of key parameters and the selection of methods. First, one of the most important parameters to set before segmentation is the segmentation scale. It is known that the optimal scale for segmentation may be different in different applications. Moller et al. (2007) proposed an approach to determine the optimal segmentation scale based on trial-and-error tests and the comparison index (Moller et al. 2007). However, there are no widely used and accepted methods that can determine the optimal scale for various applications (Myint et al. 2011). One of the most frequently used segmentation methods is the multiresolution segmentation implemented in *Definiens*, where the segmentation scale should be provided by users (Myint et al. 2011). Secondly, the feature extraction and feature selection of objects. There are a number of features that can be extracted from the objects segmented in the first step. How to select the features that are effective for accurate classification is very important. In the software *Definiens*, all kinds of object features such as texture features, geometry features, position features, and hierarchy features can be calculated. However, which features are effective and which features may be negative for classification require more studies in various applications.

2.2.7 Multiprocess Classification Model

The multiprocess classification model (MPCM) was proposed to improve the mapping accuracy of impervious surfaces at the pixel level (Luo and Mountrakis 2010; Mountrakis and Luo 2011). MPCM is actually a hybrid multiprocess model, which generally includes two classification processes, *a priori* classification and *a posteriori* classification. The general idea is to incorporate intermediate inputs extracted from the *a priori* classification results, such as linear features like roads and distance-based features. These intermediate features can be used to enhance the spectral information in optical remote sensing images to improve the *a posteriori* classification of impervious surfaces.

The successful application of MPCM model depends on the design of the *a priori* classifier, the intermediate feature extraction from the *a priori* classification, and the *a posteriori* classifier. Firstly, the *a priori* and *a posteriori* classifiers can be selected from existing classifiers such as ANNs. In previous studies, the MLP was chosen as the *a priori* and *a posteriori* classifier (Luo and Mountrakis 2010; Mountrakis and Luo 2011). Second, intermediate features include texture-based features, distance-based features, and road-based features. In the research of Luo and Mountrakis (2010), the GLCM-based texture measures were extracted from the *a priori* impervious surface classification results. Distance between nonimpervious surface and impervious surface pixels were calculated as another feature. Line-shape features were identified as road-based features. All these features were treated as intermediate inputs (IIs) to the *a posteriori* classifier. In their later study (Mountrakis and Luo 2011), the extraction of IIs was improved. For instance, road structures were further analyzed to calculate road segment properties such as segment length and directions. The texture features were extracted using directional dilation with morphological operations. Their research showed that the incorporation of IIs was able to increase the mapping accuracy of impervious surfaces by more than 3% (Luo and Mountrakis 2010; Mountrakis and Luo 2011).

2.3 ISE Using Multiple Data Sources

2.3.1 Optical and Lidar Data

Lidar has had wide applications in urban remote sensing in recent years due to its unique advantages. One of the unique features of lidar is its point cloud about elevation data. The other advantage comes from its very high resolution, generally higher than 0.5 m, since lidar data was currently obtained by aerial plane, which technically produce higher spatial resolution than satellites. A number of studies have been conducted where lidar data was frequently used to extract urban road networks, buildings, and tree structures with high accuracy (Clode et al. 2007; Elberink and Vosselman 2009; Lee et al. 2008; Miliaresis and Kokkas 2007; Tiwari et al. 2009). From our previous discussion about the MPCM approach for impervious surface estimation, it is easy to understand that these road networks, buildings, and trees can be incorporated into optical remote sensing data to improve impervious surface mapping. Multispectral aerial photograph and lidar data were combined to extract impervious surfaces (Germaine and Hung 2011; Hodgson et al. 2003). Previous studies showed that the additional use of lidar data was able to increase overall accuracy by 3% and the Kappa coefficient by 5.9%. It indicated that the improvement mainly came from the more accurate detection of buildings and trees.

2.3.2 Optical and SAR Data Using Random Forest

SAR remote sensing works in all-weather, all-day conditions, and thus is gaining an increasingly wide range of applications in different fields. SAR images can provide useful information about urban areas by reflecting the surface characteristics of urban features (Calabresi 1996; Henderson and Xia 1997; Soergel 2010; Tison et al. 2004). Various approaches are used to extract information about urban areas (Corbane et al. 2009; Dekker 2003) using both high and medium resolutions of SAR images (Dell'Acqua and Gamba 2003). These approaches include both pixel-by-pixel analysis and segmentation analysis that focused on the textural features of SAR images, and it is reported that segmentation approaches are favored over pixel-by-pixel approaches (Lombardo et al. 2003; Stasolla and Gamba 2008). Furthermore, methods of textural analysis for segmentation approaches have been comprehensively summarized in the literature (Dekker 2003; Stasolla and Gamba 2008). However, extracting urban information from SAR images remains a difficult task because of the geometric perturbations and speckle noises (Tupin and Roux 2003).

In general, fusion methods of multiple data sources (including remotely sensed data in urban areas) can be performed on three different levels: the pixel level, feature level, and decision level. Pixel-level fusion is not appropriate for SAR images, due to the existence of speckle noises. In feature-level fusion, several approaches have been proposed, including layer-stacking and ensemble-learning methods (e.g., bagging, boosting, AdaBoost, and random forest [RF] [Hall and Llinas 1997; Rokach 2010]). The ensemble-learning methods can be combined with different classifiers (e.g., ANN and SVM [Rokach 2010]). For decision-level fusion, various weighting methods (e.g., majority voting, entropy weighting, and performance weighting) and the Dempster-Shafer theory have been applied. However, conventional classifiers with a layer-stacking technique are not appropriate in this case because optical reflectance and SAR backscattering data do not correlate (Zhang et al. 2010). Among these methods, the decision tree (DT) method will be given more attention, while RF has been reported to perform very well in the fusion of optical and SAR data (Waske and van der Linden 2008).

The general strategy of RF was proposed by (Breiman 2001), which is based on randomly resampling the input training data. RF has been applied in diverse remote sensing studies (Gislason et al. 2006; Ham et al. 2005; Pal 2005), and is proven to have comparable performance to more complex methods such as SVM, which is much more time-consuming (Chehata et al. 2009; Guo et al. 2011; Waske et al. 2009). Several advantages make RF suitable for remote sensing studies (Guo et al. 2011; Yu et al. 2011). First, RF does not overfit when the number of trees increases (Breiman 2001). Second, RF does not need any feature selection since a random selection of features are built in (Yu et al. 2011). Third, RF makes no distributional assumptions about the datasets and can handle situations where the training dataset is small while the predicted dataset is large (Cutler et al. 2007).

The basic idea of RF is to grow multiple decision trees on the random subsets of the training data and related variables (Stumpf and Kerle 2011). A brief description of the RF algorithm is as follows:

Input: N training samples, with M variables/features in each sample, and μ is the size of subset of training samples, $\mu < N$.
Output: A trained RF with T decision trees:

1. Choose a training subset for a tree with replacement by T iterations in all training samples.
2. For each node, randomly choose m variables to determine the decision rules at that node. Calculate the best split based on these m variables in the training subset ($m < M$).
3. Return to Step 1 until T iterations end.

During the first step, about one-third of training samples are left out by the random selection; these samples are called out-of-bag (OOB) samples (Yu et al. 2011). The out-of-bag samples are used as the testing data for the grown decision tree.

In this study, one training sample corresponds with a location of a pixel, with the M variables/features representing the feature information from both optical and SAR images. In particular, a training sample in this case can be expressed as a vector consisting of the following two components (two groups of features): (1) reflectance of each optical band and (2) texture features of the ASAR images. The first group is from optical images and the second is derived from SAR data. For the texture features of the ASAR image, the GLCM is applied to extract the texture features (Haralick et al. 1973).

The success of RF depends on the prediction accuracy of each decision tree and the correlation between different decision trees (Breiman 2001). In order to reduce correlation, two random selection procedures are employed (Yu et al. 2011): (1) a random selection of train samples in each of the T iterations to grow each decision tree and (2) a random selection of features to select *m* features to determine each node in a tree. Therefore, two parameters are significant for the success of the RF in the fusion of optical and SAR images: the number of decision trees (*T*) and the number of features (*m*) at each node for spilling.

For the number of trees, it was reported that *T* can be any value defined by the user (Pal 2005). For the number of features, previous studies suggested *m* to be the root of the total number of features (Gislason et al. 2006; Stumpf and Kerle 2011). However, in previous applications of RF, there were often a number of features, while the number of features in this case for the ISE is very limited. Thus, it is still not clear what the optimal number should be for the random selection of features. Therefore, in this book, a quantitative analysis is designed to test the impacts of *T* and *m* on the performance of RF for the fusion of optical and SAR data for ISE.

In addition, the splitting rule is also important for the selection of features. There are several selection approaches in the literature, such as Quinlan's information gain ratio (Quinlan 1986), the Gini index (Breiman 1984), and Mingers' G statistic (Mingers 1989). The Gini index is the most frequently used for RF as it measures the impurity of an attribute by searching the largest class and isolating it from the rest of data (Breiman 1984; Pal 2003). In this book, the Gini Index is employed to measure the impurity for each node to find the best combination of features (variables). The following equation describes the Gini index of note *t* (Zambon et al. 2006):

$$\text{Gini}(t) = \sum_{i=1}^{L} p_i(1 - p_i) \tag{2.17}$$

where p_i denotes the relative frequency of each class in the training subset and *L* is the total number of classes. p_i can be determined by dividing the total number of samples of the class *i* by the total number of samples in the subset.

In this chapter, the determination of parameters, both *T* and *m* are validated and optimized for the application to optical SAR fusion in terms of ISE.

2.4 Conclusion

This chapter reviewed the literature related to research on impervious surfaces. First, the significance of impervious surfaces was introduced by summarizing their influence on the environment and socioeconomic studies. Focus was given to the hydrological impacts and atmospheric impacts of urban impervious surfaces. Second, the climatic and phenology characteristics of tropical and subtropical regions were briefly reviewed to show some of its unique features that are different from those in the temperate regions many previous studies were focused on. Third, an overview of land cover diversity was also presented to describe the impacts of the rapid urbanization process. Finally, a technical review was presented to summarize the ISE approaches using remote sensing technology. Both subpixel and per-pixel approaches were reviewed. As the per-pixel approach will be used in this book, a more detailed introduction was given to the per-pixel classification techniques with a focus on the feature extraction approaches of remote sensing images. Moreover, previous research on the synergistic use of optical and SAR data were also reviewed.

Generally, thanks to the intensive research efforts of the last decade, the significance of urban impervious surfaces has been widely recognized and assessed in numerous studies related to many urban regions, with numerous approaches proposed. However, most of this research focused on the mid-latitude cities where many important urbanization events have happened, despite the fact that with the rapid development of the global economy, more and more metropolitans are appearing in low-latitude zones such as the subtropical regions (e.g., South China, South America, South Africa, and the eastern part of Australia), where the climatology and phenology are very different from those in old urbanized areas. Moreover, the urbanization process in these newly developed metropolitans is often dramatic and diverse, making the urban land surface even more diverse and complex. Consequently, whether existing approaches are still effective for ISE in urban areas in tropical and subtropical regions becomes unclear given the special characteristics of the natural conditions and humid activities. More important, what kind of methodology is suitable for estimating the urban impervious surfaces in these tropical and subtropical regions? This book focuses on these investigations and tries to find appropriate solutions to address these issues.

3

Methodology of Combining Optical and SAR Data

3.1 Study Area

Six cities from four countries located on the tropical and subtropical areas (Figure 3.1) were carefully chosen as the study sites of this research. These cities are Guangzhou, Shenzhen, Hong Kong, Sao Paulo, Mumbai, and Cape Town. A basic description about these six cities is given in the following sections including their geography, climate, population, economy, and urban planning.

3.1.1 Site A: Guangzhou

Guangzhou, the capital city of the Guangdong province, China, is located at the center of the PRD metropolitan. The PRD is located downstream of the Pearl River, and is known as the third largest metropolitan in China, enjoying a tremendously fast development during the past 30 years. The region has had a quick urbanization process, with a population of over 19 million and an area of over 21,000 km^2 (Fan et al. 2008). However, due to strong interactions between human activities and the environment, serious environmental issues have also emerged and are causing a series of environmental problems, including air pollution and water pollution (Zhang et al. 2008). Nevertheless, unlike many other metropolitans in the world, the PRD is located in a subtropical humid area, with long period of cloudy weather throughout the whole year and different characteristics of plant phenology (Fan et al. 2008). In the PRD, about 80% of precipitation occurs during April to September, which is known as the wet season, while only about 20% of rainwater occurs from October to the following March, which is known as the dry season (Cai et al. 2004). Therefore, the seasonal effects of ISE from remote sensing imagery are likely to be different from those found in previous research, which was conducted mainly in midlatitude regions. Three sites in the PRD, Guangzhou, Shenzhen, and Hong Kong, have been chosen as the study sites in this book, with detailed descriptions in the following sections.

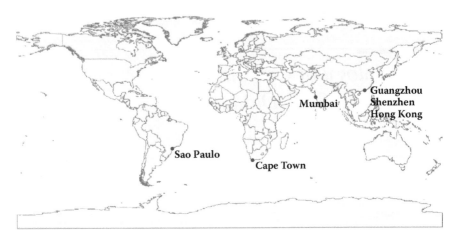

FIGURE 3.1
Locations of the six cities from four countries in this study.

Guangzhou has undergone a dramatic urbanization process and is the third largest city in China. The population of Guangzhou reached 12.78 million in 2010 (Brinkhoff 2011). The study site selected in this research is located in the Huangpu District of Guangzhou (Figure 3.2), which is a medium urbanized area. The land cover in this site is characterized by residential areas, small rivers, farmland, small hills, and small lakes and water pools with seasonal waters. Impervious surfaces can be seen from the Landsat ETM+ image with both high and low reflectance. Within this study, this site was selected to evaluate ISE using Landsat ETM+ and ENVISAT ASAR

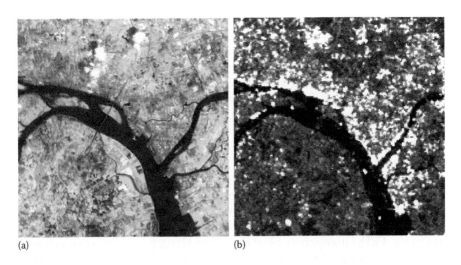

(a)   (b)

FIGURE 3.2
(a) Landsat ETM+ (RGB: 5-4-3) and (b) ENVISAT ASAR images of the study site in Guangzhou.

images, as well as to investigate the seasonal effects of the plant phenology and climate of the tropical and subtropical regions.

3.1.2 Site B: Shenzhen

Shenzhen is located in the southern part of the PRD and to the north of Hong Kong. Shenzhen is China's first and most successful special economic zone (SEZ), which has been highly urbanized in the past three decades. Commercial and industrial areas are intensively distributed all over the city due to the rapid development of the economy (Figure 3.3). The selected site is on the boundary of Shenzhen and Hong Kong, which is highly urbanized and is characterized with mainly commercial, residential, and greening areas of the city. SPOT-5 and ENVISAT ASAR images in Shenzhen were selected to test the effectiveness of the methods proposed in Section 3.5.

3.1.3 Site C: Hong Kong

Hong Kong is situated in the southern part of the PRD on the coast of the South China Sea. It consists of three main parts, the New Territories, Kowloon, and Hong Kong Island. Even though Hong Kong has been intensively urbanized, it is a mountainous city with a large area of mountains distributed all over the city. The impervious surface percentage can reach up to 100% in urban areas such as Kowloon, but is only moderately urbanized with a moderate impervious surface percentage in the rural areas such as the New Territories. In this study, a site located in the Yuen Long in the northern part of the New Territories was selected (Figure 3.4). This study site is typical in Hong Kong

(a)

(b)

FIGURE 3.3
(a) SPOT-5 (RGB: 4-1-2) and (b) ENVISAT ASAR images of the study site located in Shenzhen.

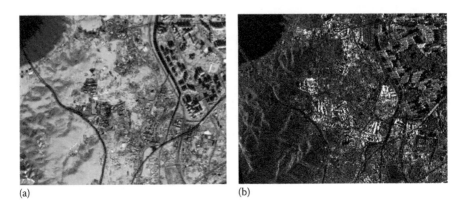

(a)   (b)

FIGURE 3.4
(a) SPOT-5 (RGB: 3-1-2) and (b) TerraSAR-X images of the study site in Hong Kong.

as it is moderately urbanized and includes both plain and mountain areas. Land covers in this study site include residential areas, farmland, mountains, and coastal sea surface. SPOT-5 and TerraSAR-X images in this area are used to address the ISE and evaluate comprehensively the effectiveness and performance of the synergistic use of optical and SAR data.

3.1.4 Site D: Sao Paulo

Sao Paulo is located in southeastern Brazil and is the largest city in Brazil by population and gross domestic product (GDP). Sao Paulo has a humid subtropical climate and is significantly influenced by monsoons, and the average annual precipitation is about 1454 mm. The Sao Paulo metropolitan area

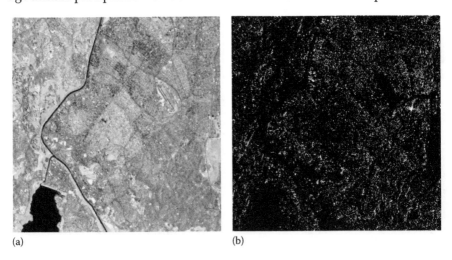

(a)   (b)

FIGURE 3.5
(a) Landsat TM (RGB: 5-4-3) and (b) ENVISAT ASAR images of the study site in Sao Paulo.

has been undergoing a rapid urbanization process since the twentienth century. However, urbanization has brought significant environmental impacts to the ecosystem, such as the deforestation of rainforest (Torres et al. 2007). Therefore, the estimation of impervious surfaces would be beneficial for urban planning and environmental management of the city. In this study, Landsat TM, ENVISAT ASAR, and TerraSAR-X are used synergistically to extract the impervious surfaces of Sao Paulo. Figure 3.5 shows the Landsat TM and ENVISAT ASAR data in this study area.

3.1.5 Site E: Mumbai

Mumbai is located in western India, with a total urban area of approximately 465 km². Mumbai has a tropical wet and dry climate, with an average annual precipitation of about 2167 mm and an average annual temperature of 27.2°C. It is the main city of western India and a leading economic and financial center (Bhagat 2011; Moghadam and Helbich 2013). The population of Mumbai has nearly doubled in the last four decades according to the Indian Census of 2011 (Moghadam and Helbich 2013), with an increasing trend to reach about 27 million by 2025 (United Nations 2012). However, there are many problems introduced by the rapid urbanization in Mumbai, such as urban fragmentation (Gandy 2008). Therefore, remote sensing of the urbanization process of Mumbai would be very helpful to monitor the urban sprawl in order to improve the urban planning and management of Mumbai. In this study, Landsat TM, ENVISAT ASAR, and TerraSAR-X are used to estimate the impervious surface distribution of Mumbai. Figure 3.6 shows the Landsat TM and ENVISAT ASAR data in this study area.

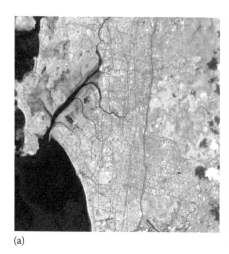

(a)

(b)

FIGURE 3.6
(a) Landsat TM (RGB: 5-4-3) and (b) ENVISAT ASAR images of the study site in Mumbai.

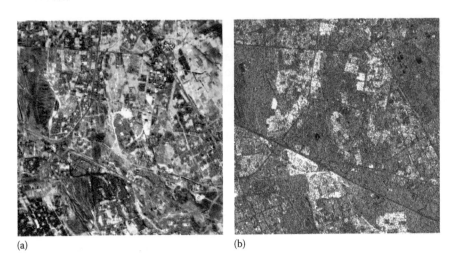

(a)   (b)

FIGURE 3.7
(a) Landsat TM (RGB: 5-4-3) and (b) ENVISAT ASAR images of the study site in Cape Town.

3.1.6 Site F: Cape Town

The city of Cape Town is located in the southwestern part of South Africa, covering about 2460 km². It is located at approximately latitude 33.55°S and longitude 18.25°E, which is nearly on the boundary of the subtropical region in the Southern Hemisphere. Cape Town enjoys a Mediterranean climate with warm and dry summers and cool and wet winters. The population of Cape Town was about 3.5 million in 2008 (Rebelo et al. 2011). Cape Town has been undergoing a rapid urbanization process with significant land use/land cover changes (Rebelo et al. 2011). Consequently, this rapid urbanization has caused great environmental impacts, especially damage to the biodiversity (Rebelo et al. 2011). Satellite monitoring of the urban sprawl in Cape Town will be important in order to monitor these impacts in a timely manner. In this study, Landsat TM, ENVISAT ASAR, and TerraSAR-X will be used to estimate the impervious surface distribution of Cape Town. Figure 3.7 shows the Landsat TM and ENVISAT ASAR data in this study area.

3.2 Satellite Data

3.2.1 Landsat ETM+

The Landsat ETM+ images had one panchromatic band and six bands with an image pixel size of 30 m × 30 m. In this study, only the 30 m data was used. The ETM+ image was acquired on December 31, 2010. In order to preprocess ETM+ images, a process should first be applied to get rid of the stripes on the eastern and western edges of each scene that are caused by the footprints (location and spatial extent) of each band due to scan line corrector (SLC) failure. For this reason, the study area was located in the middle of each scene where there were no stripes, and thus no stripe removal operation was applied. We assumed that the atmospheric conditions were clear and homogeneous and the small area of clouds would not significantly impact the whole scene of the image, and thus no atmospheric correction was performed (Wu and Murray 2003).

3.2.2 SPOT-5

SPOT is a high-resolution optical satellite family launched by France. SPOT-5 was launched on May 4, 2002, with a higher spatial resolution of 2.5 and 5 m in panchromatic mode, and 10 m in multispectral mode. The SPOT-5 image used in this study was in a multispectral mode image, at precision 2A level, and was obtained on November 21, 2008. Therefore, the pixel size of the SPOT-5 data in this study is 10×10 m. The multispectral SPOT-5 data has four image bands located in the green region (500–590 nm), red region (610–680 nm), near-infrared region (780–890 nm), and shortwave infrared region (1580–1750 nm). The image was projected under the coordinate system of World Geodetic System 1984 (WGS84) and Universal Transverse Mercator (UTM) (Zone 50N).

3.2.3 ENVISAT ASAR

ASAR is a radar instrument on the ENVISAT satellite operated by the European Space Agency (ESA). ENVISAT was launched on March 1, 2002 with a projected mission duration of 5 years and continued to work for 10 years. Even though it stopped operating on April 8, 2012, the archive data is still beneficial for this study due to the nature of the study and the selected time period. ASAR operates in the C band (4–8 GHz) and generally has five operation modes: alternating polarization (AP) mode, image (IM) mode, wave (WV) mode, suivi global (GM) mode, and wide swath (WS) mode, where AP, IM, WV, GM, and WS are the identity codes. Raw data from these operation modes is the Level 0 data, and they can be further processed to Level 1 or even higher levels of data product by different treatments. The ASAR data used in this study is the wide swath mode (WSM) and IM precision (IMP) data, which is a Level 1b data product. The data was received by the Satellite Remote Sensing Receiving Station at the Chinese University of Hong Kong. The ASAR WSM data was obtained on September 23, 2010, on the descending direction with vertical transmit/vertical receive (V/V) polarization and a pixel size of 75×75 m. The ASAR IMP data was obtained on November 19, 2008, on the ascending direction, Track-25 of ENVISAT, with V/V polarization and a pixel size of 12.5×12.5 m. Additionally, due to the uncertainty of speckle noises in SAR images, the enhanced Lee filter is selected to filter the speckle noises. The enhanced Lee filter is an improved version of the Lee filter that was designed to better preserve texture information, edges, linear features, and point targets in SAR images (Lee 1983). The enhanced Lee filter is an adaptive filter that was proven to be more suitable for preserving radiometric and textural information than other speckle filters (Lopes et al. 1990; Xie et al. 2002). The ASAR image was then geocoded and projected with the georeference system of the WGS84 and UTM (Zone 50N).

3.2.4 TerraSAR-X

TerraSAR-X (TSX) is a German earth observation satellite launched on June 15, 2007 and is still in operation. TSX operates in the X band (9.6 GHz) and has three main imaging modes: SpotLight mode, StripMap mode, and ScanSAR mode. The TSX image used in this study is in StripMap mode, obtained on November 16, 2008, with a spatial resolution of 3×3 m, and the scene size is 30 km (width) $\times$ 50 km (length). The TSX image was geocoded with Next ESA SAR Toolbox (NEST) 4C-1.1 software developed by ESA under the coordinate system of WGS84 and UTM (Zone 50N). Geometric correction was also conducted by the Range-Doppler Terrain Correction in NEST with digital elevation model (DEM) data. Additionally, due to the uncertainty of speckle noises in SAR images, the enhanced Lee filter is selected to filter the speckle noises.

3.3 Digital Orthophoto Data

The orthophoto is derived from aerial photographs that were taken mainly at a flying height of 2400 m in Hong Kong, and were named the Digital Orthophoto DOP 5000 series. The whole land area of the Hong Kong Special Administrative Region (HKSAR) is covered by 190 tiles of DOP 5000 images with a specific tile number for each image. The original DOP 5000 data has a ground pixel size of 0.5×0.5 m and is georeferenced in the coordinate system of the Hong Kong 1980 Grid. The DOP 5000 photo used in this study was taken on November 12, 2008, and was located on the northwestern part of Hong Kong with a tile number of 6-NW-A. To use the DOP 5000 data in this study, a coordinate system transformation was conducted on the data to transform it from Hong Kong 1980 Grid to UTM50N with WGS84.

3.4 *In Situ* Data

Field work was conducted on January 7, 2013 (winter) to collect information about the spectral reflectance and spatial texture of different land covers.

The field work was carried out in the Yuen Long district located in north-western Hong Kong (Figure 3.8).

A Global Positioning System (GPS) device (Leica Zeno) was employed to locate the geocoordinates of the fields (Figure 3.9). Using a Nokia cell phone (Nokia 5320) to connect to the local differential GPS (DGPS) reference station during the field work, the DGPS technique was used to improve the location accuracy, which was up to 0.4 m. A spectrometer produced by Analytical Spectral Devices, Inc. (ASD) was employed to collect the hyperspectral reflectance of each land cover. The field of view (FOV) of the ASD spectrometer is 25 degrees and the area to be measured was set as 0.5 × 0.5 m according to the resolution of the digital orthophoto data, and therefore, the height of the ASD spectrometer should be 1.13 m (= 0.25/tan12.5°) from the ground. Moreover, a digital camera (Canon-Digital IXUS 115 HS) was used to take

FIGURE 3.8
Image from Google Earth showing the area where field data was collected.

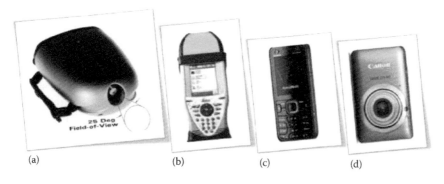

(a)   (b)   (c)   (d)

FIGURE 3.9
Devices employed for field data collection: (a) ASD spectrometer, (b) Leica Zeno, (c) Nokia cell phone, and (d) Canon camera.

photos of each land cover, which were used to help analyze the color, texture, and shape features. During the field work, six main land cover types were considered, including dark impervious surfaces (i.e., asphalt road and old concrete roads), bright impervious surfaces (i.e., new concrete roads), vegetation (i.e., shrubs and grasses), farmland (i.e., crop), water surfaces (i.e., rivers and water pools), and bare soil in a field under construction. In this book, the collected field data were used to validate the results of visual interpretation of the satellite images.

3.5 Framework of Methods

Figure 3.10 shows the framework of methodology of this study, illustrating the mechanism of land cover diversity, the responses of both optical and SAR remote sensing, and the ISE from remote sensing images. The main idea shown in Figure 3.10 is that the difficulties of ISE from satellite images are caused by the diversity of land covers and their reflectance in the images. Moreover, land cover diversity is caused by the phenology and the climatology of tropical and subtropical regions and the extreme human activities surrounding urbanization. Meanwhile, the phenology is affected by the climatology, which mainly includes the characteristics of temperature and precipitation.

Therefore, the methodology of this study includes three parts: (1) to investigate the effects of phenology and climatology on ISE, aiming at finding the most suitable season for ISE from satellite images, (2) to investigate the characteristics of urban land covers, which are the direct cause of the difficulties in the accurate estimation of impervious surfaces, and (3) to address the methods for synergizing optical and SAR images in order to improve the accuracy of ISE.

3.5.1 Per-Pixel Modeling of Impervious Surfaces

Impervious surface mapping at a per-pixel level is actually a classification process where impervious and nonimpervious surfaces are a combination of various land cover types. Conventional LULC includes vegetation, urban areas, and water, and each land cover type shares similar spectral and spatial characteristics. Therefore, they are often identified individually during the classification procedure. However, impervious and nonimpervious surfaces consist of various land cover materials. For instance, impervious surfaces can be made up of dark materials (e.g., asphalt and old concrete) and bright materials (e.g., new concrete and metal), while nonimpervious surfaces are also very diverse in materials (e.g., vegetation, water, and base soils). In this study, a two-step approach is employed to estimate the impervious surfaces. First, six land cover types: dark impervious surfaces, bright impervious

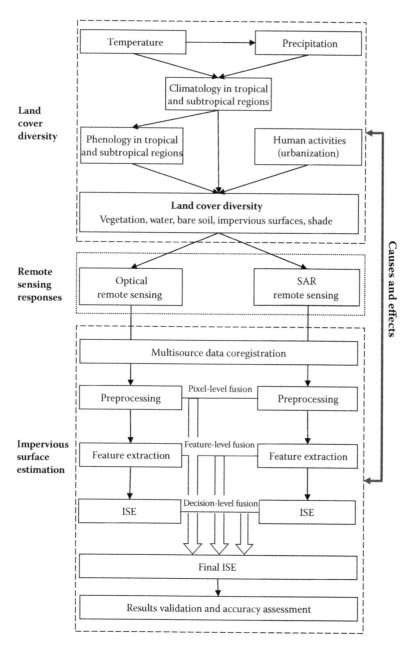

FIGURE 3.10
Framework of the methodology of this research.

surfaces, vegetation, water body, bare soil, and shaded areas, are identified with a classification procedure using RF. Second, a procedure is conducted to combine various land covers into impervious and nonimpervious surfaces.

In particular, shaded areas are treated as a single land cover type as they often have unique spectral and spatial characteristics. Moreover, since shaded areas may be impervious (e.g., roads and rooftops) or nonimpervious (e.g., greening areas), they are treated as nonimpervious surfaces in the second step of combination in this study. Therefore, dark impervious surfaces and bright impervious surfaces are combined as impervious surfaces and vegetation, water, bare soil, and shade are combined as nonimpervious surfaces. Additionally, because misclassification may happen not only between impervious and nonimpervious land cover types but also among different subtypes of impervious or nonimpervious types, the accuracy of classification before and after the combination operation may be different. Therefore, in this study, an accuracy assessment is conducted on the classification results before and after combining impervious surfaces and nonimpervious surfaces subtypes.

3.5.2 Investigation of Seasonal Effects

The motivation of this study is to prove or reject two hypotheses. First, seasonal changes of the landscape components should be less problematic in subtropical monsoon areas since vegetation and canopy change less among the seasons, while water may have an impact, as there are many variable source areas (VSAs). VSAs are those areas filled with water in rainy seasons, and bare soil is exposed in dry seasons (Frankenberger et al. 1999). Second, the sensitivity of the ISE may depend on different methods. In this book, the per-pixel approach is adopted and two popular classifiers are selected, including ANN and SVM.

This book investigates the variation of ISE from different seasons of satellite images and the seasonal sensitivity of different methods. Four Landsat ETM+ images of four different seasons are employed to estimate the impervious surfaces at the pixel level. Seven land use types are defined to conduct the classification procedure according to the landscape of the study area. Table 3.1 gives a brief description about each land cover type, including water, vegetation, bare soil, clouds, shade, dark impervious surfaces, and bright impervious surfaces. In particular, clouds and shade are treated as one type of land cover, since they have unique spectral and spatial characteristics compared with other types of land cover. As the region is undergoing a dramatic urbanization process, a lot of bare soils appear on the areas under construction. Further, numerous cool roofing materials, which are light blue or white in color, are used to build up rooftops. These rooftops are designed to highly reflect the solar radiation in order to reduce the urban heat island effect. Thus, they appear to be bright impervious surfaces in

TABLE 3.1

Definition of the Land Covers Used in This Study

Land Use Type	Definition
Water	Rivers, lakes, and other freshwater bodies
Vegetation	Grain crops, vegetable crops, grass, and other agricultural land
Bare soil	Land under construction with bare soils exposed
Clouds	Small and fragmentary clouds that are difficult to remove
Shades	Topographical shades and shades from tall buildings, trees, etc.
Dark impervious surfaces	Rooftops, roads, and parking lots that are made of asphalt, concrete tile, and other materials with low spectral reflectance
Bright impervious surfaces	Cool rooftops and green rooftops that are made of cool materials, such as metal, which are designed to highly reflect solar radiation

Note: Clouds and shade are treated as land covers since they have unique spectral and spatial characteristics compared with other land covers in satellite images.

the optical remote sensing data. ANN and SVM are then applied to classify the four seasonal changes of ETM+ images. After seven land use classes are available, a combining operation is employed to reclassify the five land use types into two types: the impervious surfaces and nonimpervious surfaces. During this period, the water body, vegetation cover, and bare soils are combined together to form the nonimpervious surfaces, while the dark and bright impervious surfaces are combined into the impervious surfaces class.

3.5.3 Feature Extraction

3.5.3.1 Conventional Feature Extraction

According to previous literature, segmentation methods are superior over pixel-by-pixel methods because segmentation methods take texture characteristics into account (Dell'Acqua and Gamba 2003; Stasolla and Gamba 2008). A texture characteristic is important for the interpretation of SAR data because the speckles in SAR data result in difficulties for the pixel-by-pixel approaches. Therefore, in order to extract complementary information for urban impervious surfaces from ASAR data, texture feature extraction is necessary and important. In this study, the popular GLCM approach (Haralick et al. 1973) is employed to analyze the texture features of the ASAR data. For the application of GLCM, the size of image block and the texture measures with GLCM have been a major issue (Marceau et al. 1990). In terms of the classification of remote sensing images in urban areas, it is reported that a window size of 7 × 7 pixels is suitable with a test on the resolutions from 2.5 × 2.5 m to 10 × 10 m (Puissant et al. 2005). Moreover, four texture measures: homogeneity (HOM), dissimilarity (DISS), entropy (ENT), and angular second moment (ASM) were identified as effective indicators for the texture description of different urban land cover types (Puissant et al. 2005). Thus, in this study, the window size is set as 7 × 7 pixels, and four texture measures, HOM, DISS, ENT, and ASM, are employed.

3.5.3.2 SAN Feature Extraction

3.5.3.2.1 Concept of Shape-Adaptive Neighborhood

The neighborhood is a basic and key concept in image processing. However, previous feature extraction approaches using neighborhoods with regular shapes had some shortcomings that could lead to error because terrain objects may have different irregular shapes. As more attention has been paid to human cognition, the procedure of human vision has been considered and applied in image processing. Considering human vision, the color and shape of the target are very important. Human beings recognize different objects by their color characteristics first, then by their shape feature, and other features such as texture. This procedure of recognizing an object generally happens within a local neighborhood in the image. If the object is gray, then the shape characteristics will be the most important features for human eyes because there are no colors with which to identify different objects. Based on this observation, the concept of SAN was proposed to start the procedure of feature extraction. Prior to the feature extracting procedure, color characteristics will be analyzed to determine the neighborhood of each pixel with an adaptive boundary.

Definition

A SAN is the neighborhood of a pixel containing but not necessarily centered on the pixel, whose shape is determined by the terrain object it represents (Zhang et al. 2013). ■

Figure 3.11 shows the concept of the SAN of a pixel (a), where the view port is used to represent the local range to search the SAN or the object. The

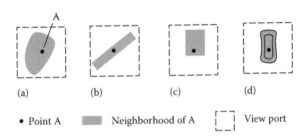

(a)   (b)   (c)   (d)

• Point A   ▬ Neighborhood of A   ⌐ ⌐ View port

FIGURE 3.11
Illustration of a SAN of point A (a) in an irregular shape object, (b) in a rectangular object, (c) in a square object and (d) in a complex environment (e.g., sports ground).

feature of a SAN only represents the feature of the central pixel (not always in the center). If the pixels in the SAN are of the same terrain object as the central pixel, the judgment is correct. In contrast, if there are some misjudged pixels in the SAN, this will affect only the classification of the central pixel without any impact on other pixels in the SAN.

After determining the SAN of a pixel, the feature of the SAN can be extracted, including the color, texture, and shape feature, which describe characteristics of the central pixel and will then be used in the classifying procedure. The mechanism of the determination of SAN is consistent with that of the on-off switch model in the attention mechanism of human vision (Solso et al. 2004). According to cognitive psychology research, in a local range containing an object of the on-off switch model, those parts that do not belong to the object will be filtered out and thus do not catch our attention. Only the object itself will be able to pass the on-off switch and then cause the so-called attention. This phenomenon has also been proven with some evidence in neural experiments both in animals and humans (Solso et al. 2004). It is important to note that the SAN here is only suitable in those objects with no significant variation in color, or in the case of remote sensing images, with no significant variation in multispectral reflectance. However, a certain extent of variation is allowed because there is little variance in spectral reflectance (color), and this is the common case in reality. This situation can be handled with a heterogeneity-based threshold, which will be discussed in the following section.

3.5.3.2.2 Determination of a SAN

3.5.3.2.2.1 Spectral Feature Transformation Multiple spectral features are crucial for the interpretation of remote sensing images. However, the problem of how to understand the information contained in different spectral bands becomes apparent when using visual interpretation, since a visible color (for human eyes) consists of only three components in existing popular color space (e.g., red, green, blue [RGB] and HSI). In this case, three of the bands are often assigned to be the red, green, and blue values used to generate a false-color image for visual interpretation of the image. Moreover, different combinations of bands are attempted in order to discriminate between different objects. This RGB mode of mapping the spectral feature space is also called the color space, referred to as the color characteristics in visual interpretation.

As discussed above, determination of the SAN depends on the color characteristics. Heterogeneity is used to describe the color feature as follows. Conventionally, there are several color models, including RGB, hue, saturation, intensity (HSI), and hue, saturation, value (HSV). In the processing of remote sensing images, the images are often transformed to the false-color composition; that is, in the RGB color space. However, according to the existing literature, the RGB color space is not consistent with human vision (Herodotou et al. 1999). A color point in RGB color space cannot really represent the color recognized by human eyes as it corresponds differently from that of the human perception of color. Of these color spaces, the one closest to the human perception of color is HSV (Herodotou et al. 1999). The transformation formula from RGB color to HSV color is shown in Equation 3.1 (Herodotou et al. 1999), where the value of H would be in the range [0, 360], and the values of S and V would be in [0, 1].

$$H = \begin{cases} \cos^{-1}\left\{ \dfrac{(R-G)+(R-B)}{2\sqrt{(R-G)^2+(R-B)(G-B)}} \right\} & \text{if}(B \leq G) \\ 360° - \cos^{-1}\left\{ \dfrac{(R-G)+(R-B)}{2\sqrt{(R-G)^2+(R-B)(G-B)}} \right\} & \text{if}(B > G) \end{cases}$$

(3.1)

$$S = \frac{\max(R,G,B) - \min(R,G,B)}{\max(R,G,B)}$$

$$V = \frac{\max(R,G,B)}{255}$$

In order to treat the three components in the same way for the calculation of the color feature, the H needs to be transformed in the range [0, 1]. For the hue component, the visible spectrum distributes over the whole range [0, 360]; that is, including 0 but excluding 360. For instance, red, green, and blue colors are separated by 120° within this range. In this way, H does not assume the mathematical meaning of an angle (i.e., 90° and 270° represent different colors). Thus, it is reasonable to normalize H to [0, 1], which includes 0 but excludes 1, with a linear transformation. After the transformation, the color feature of the pixel can be expressed as

$$CF = \omega_1 \cdot H + \omega_2 \cdot S + \omega_3 \cdot V$$

(3.2)

where ω_1, ω_2, and ω_3 are the weights of the three components, and $\omega_1 + \omega_2 + \omega_3 = 1$. Therefore, the color feature CF here will be a single value instead of a vector of the three components. There are two advantages of this. First, it is convenient to place different weights with different components. According to psychological research, hue is related mostly to the color we determine (see) an object to be. Thus, it is reasonable to place higher weight on the H component; that is, ω_1. However, which combination of the three weights is the best probably depends on various applications. The second advantage is that it is computationally better than representing the color feature as a vector. Since the color feature is used to calculate the heterogeneity for a large number of times in the following steps, the representation of the color feature in this way will save a lot of time.

3.5.3.2.2.2 Determine SAN   The SAN of a pixel is determined within a view port (Figure 3.2) centered on the central pixel using a given heterogeneity threshold. The heterogeneity between two pixels is defined to determine the SAN of one pixel using its color feature. Let CF_0 be the color feature of the central pixel and CF_i represent the color feature of the pixel i, which is to be determined whether inside the SAN or not. Thus, a simple way of expressing the heterogeneity between the two pixels is $diff = |CF_0 - CF_i|$. Given a threshold T and that the SAN of the central pixel is SAN_0, the rule could be $i \in SAN_0$ *iff diff* $< T$, where *iff* represents the term "if and only if."

The threshold of heterogeneity between two pixels is a key factor influencing the size of the SAN. If the threshold is too small, most of the SANs contain only a few pixels, which can result in difficulties for the feature extraction from the SANs. Thus, an appropriate threshold is crucial for the feature extraction procedure in the steps to follow. In a simple way, the optimal threshold is found with a threshold search procedure by quantitatively testing various thresholds and analyzing the size of the SANs. In this study, a series of numbers is tested for threshold values and corresponding sizes of SANs are counted. Results are plotted in a curve using a spline interpolation method. Finally, the threshold curve is used to determine the optimal threshold.

3.5.3.2.3 Extracting Spatial Features

3.5.3.2.3.1 Texture Extraction Method   There is no accepted quantitative definition of texture (Bharati et al. 2004). Rather, it is left as an intuitively obvious but quantitatively undefined characteristic associated with a given pixel. Various attempts have been made to give it an appropriate quantitative definition, but none appear to have achieved widespread acceptance. In this study, a texture analysis is conducted on each SAN to represent the texture characteristic of each pixel. Texture features can be extracted with the SAN. There are many methods cited in the previous literature for carrying out texture analysis, such as co-occurrence-matrix-based approaches (Zhang 2001), random distribution models (Bruzzone and Prieto 2002), and geostatistical methods (Curran 1988). Since each SAN has an uncertain shape, and considering the case of remote sensing images, the geostatistical approach was used to describe the spatial autocorrelation, which is closely related to the texture characteristics (Jensen 2007). The geostatistical approach was reported to successfully represent the autocorrelation of spatial data (e.g., remote sensing images) (Fabbri et al. 1993; Jensen 2007). In geostatistics, the variogram is calculated first and is fitted with a theoretical model such as the spherical model, and then the parameters of the variogram, such as nugget, sill value, and variable-range, are used to describe the characteristics of spatial autocorrelation.

However, the calculation of a variogram and the fitting of the theoretical variogram are time-consuming processes. Since we do not employ the variogram to predict some unknown pixels, but only to describe the texture feature, there is no need to calculate all the function values of the variogram at every step length h. Only some of the key steps are helpful to describe the texture feature, such as when $h = 1$, the function value $\gamma(h)$ will be the sill value of the variogram. Thus, a selected series of steps was used to compute the function values, which is a modified version of the variogram, shown as follows:

$$\gamma(H) = \frac{1}{N(H)} \sum_{i=1}^{N(H)} [Z(x_i) - Z(x_i + H)]^2 \qquad (3.3)$$

where $H = [h_1, h_2, ..., h_n]$ denotes the selected series of steps and $\gamma(H)$ the resampled variogram, which is treated as the extracted feature of the texture. In specific applications, h should be selected according to both the spatial resolution of the data and the landscape characteristics of the land surface. Generally, the higher the resolution of the data and the larger the size of the targets, the higher the value of h should be used. In this study case, H is empirically set to be [1, 2, 3].

3.5.3.2.3.2 Description of Geometric Features and Their Effectiveness   Geometric features include many types, such as shape features and topological features. Since this study focuses on modeling the early processing of visual perception, only the shape features are considered. Shape features of images have received attention since 1993, when Fabbri first introduced the shape feature into multispectral remote sensing images (Fabbri et al. 1993; Jensen 2007). Description of shape characteristics in the traditional shape analysis methods contained the compact expression (compactness), the complexity description, and the curvature description. In this study, two kinds of compact expressions are employed as the shape descriptors: the aspect ratio (R) and the form factor (F), illustrated by Equation 3.4, where L and W are the length and width of the minimum boundary rectangle (MBR) of the SAN, B is the perimeter of the SAN, and A is the area of the SAN.

$$R = \frac{L}{W}; \quad F = \frac{|B|^2}{4\pi A} \qquad (3.4)$$

Another issue of the shape feature extraction is the effectiveness of the shape description. The shape characteristics may be meaningless for some terrain objects with random shapes, such as natural forests, residential areas, and farmland, but they are important for roads, buildings, sports fields, and other targets with a regular shape. Therefore, the effectiveness of the shape feature is very important. The effectiveness of the shape can be defined as

$$eff = [Re, Fe] \qquad (3.5)$$

where $Re \in \{0, 1\}$ is the effectiveness of aspect ratio, $Fe \in \{0, 1\}$ is the effectiveness of the form factor, and if the shape feature is meaningful for classification, it is assigned to be 1; otherwise, it will be 0. The assignment of shape effectiveness can be done with a supervised classification procedure, which can be conducted via four steps: (1) visually select a set of samples containing pixels with both effective and ineffective shape features, (2) generate the SANs of these samples and calculate their shape features, (3) use these samples to train a classifier (e.g., minimum distance classifier or maximal likelihood classifier) to classify the shape features (the output of the classification is 0 or 1), and (4) apply the classifier to assign the effectiveness of the shape features for other SANs.

Finally, the shape features of a SAN can be expressed as the Equation 3.6 in vector form. However, some managed forests and balanced residential areas may also have regular shapes. In this case, the determination of shape effectiveness should be more complicated than that presented above. For instance, texture features should be taken into account to help define the shape effectiveness. In this book, we only consider natural forest and ordinary residential areas.

$$SHA = [R, F, Re, Fe] \tag{3.6}$$

3.5.3.2.3.3 Integration of All the Features   The feature of a SAN contains the color feature, the texture feature, and the shape feature. The values of all these features are normalized into the same range [0, 1]. Finally, all features should be integrated to express the general feature of each SAN. This feature integration procedure is a data fusion procedure on the feature level, which can be illustrated by the following general model.

$$SANF = fusion(CF(k), TF(m), SF(n)) \tag{3.7}$$

where $CF(k)$ is the color feature, $TF(m)$ is the texture feature, and $SF(n)$ is the shape feature. Each feature is multidimensional and can be represented as a vector, and the variants k, m, n are the number of vector elements for color, texture, and shape features, respectively. The function *fusion()* is the fusion method used to integrate all the features. In Equation 3.7, the function *fusion()* is only a concept model. Many existing approaches can be used to conduct this operation, such as principal component analysis (PCA), Fourier transform, and wavelet transform, or simply link all three vectors of features into a larger vector if there are only a small number of features. In the experiment conducted in this study, there is one color feature (Equation 3.2), three texture features (Equation 3.3 where $H = [1, 2, 3]$), and four shape features (Equation 3.6). Thus, only the simple vector-based approach is used as there are only a total of eight features (= 1 + 3 + 4) in the study case.

3.5.4 Fusing the Optical and SAR Data

A flowchart of fusing the optical and SAR images is shown in Figure 3.12, illustrating the details of the combination of optical and SAR images to classify impervious surfaces. Several issues are important for the fusion of optical and SAR data, including the coregistration of the two data sources,

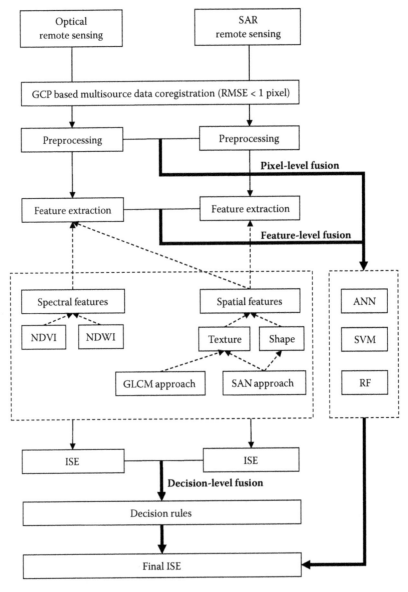

FIGURE 3.12
Optical-SAR fusion for ISE.

the feature extraction methods, the comparison between a single use of optical or SAR data, the comparison of the differences of various levels of fusion strategies, and the fusion methods for combining the two data sources. Since the feature extraction methods have already been described in Section 3.5.3, this section will focus on the methods of image coregistration, the comparison of the single use of the two data sources, the comparison of different levels of fusion, and the fusion methods between optical and SAR images.

3.5.5 Result Validation and Accuracy Assessment

For supervised classifiers (e.g., MLP, SVM, and RF), both the size and quality of the training data are highly significant in determining the performance of classification, while testing data is imperative in assessing the accuracy of both supervised and unsupervised classifications. In order to quantitatively assess the accuracy of the ISE, an appropriate sampling framework should be used. Five main types of sampling schemas were summarized previously by Jensen (2007): the simple random schema, the system schema, the layer schema, the layer system schema, and the cluster schema, where different backgrounds represent different classes in the layer schema. Different schemas are suitable for different cases and require different amounts of work. Among these schemas, the cluster schema is the most convenient to conduct, with much less labor than the other schemas. In this study, the cluster schema was applied to the study area by sampling the cluster test data over the satellite data with the aid of visual interpretation of the satellite data, digital orthophoto data, and *in situ* data. Moreover, very-high-resolution data from Google Earth is also used to help visually interpret the satellite data for the result validation in this study.

3.6 Conclusion

This chapter presented the methodology used in this book. A general framework of methodology was first introduced to explain the logical relationship of land cover diversity, remote sensing responses, and ISE. Second, the per-pixel modeling of ISE was presented as a basic strategy to estimate impervious surfaces in this research, followed by an introduction of the approaches to investigate the seasonal effects of ISE in tropical and subtropical regions. Third, the feature extraction methods were presented in detail. Following an introduction to the conventional feature extraction approach based on GLCM, a novel approach based on the SAN was presented with technical details. Fourth, a methodological framework of fusing the optical and SAR data was presented with methods of image coregistration, investigation of the advantages and disadvantages of optical and SAR data, comparison of different fusion levels, and the fusion procedure with supervised classifiers. Finally, the sampling methods for training and testing datasets and the accuracy assessment approach were presented.

4

Impact of Climate Zone on Impervious Surface Estimation and Mapping

4.1 Introduction

A climate zone can have significant impacts on impervious surface estimation using satellite images in tropical and subtropical areas. There are several major climate categories in the regions of concern, including tropical moist climate, wet-dry tropical climate, humid subtropical climate, and Mediterranean climate, in which the Mediterranean climate zone is located on the boundary between subtropical and temperate regions. Different climate zones have different seasonal patterns of temperature, precipitation, humidity, plants, and so forth. A climate zone can influence impervious surface estimation in different ways directly or indirectly. This influence is also known as the seasonal effects when using optical remote sensing images to map impervious surfaces. These seasonal effects include three aspects: (1) the difference in precipitation in dry seasons and wet seasons can produce a difference of water surface area on the land surface, (2) the seasonal changes of soil moisture in rainy and dry seasons can influence the spectral confusion between different land cover types, and (3) the seasonal changes of plants will change the vegetation coverage in the areas of hills, mountains, and greening zones in urban areas. Water body and vegetation are two important land cover types in urban remote sensing studies, and the distribution of these two land cover types lead to different patterns of spectral confusion in a given urban study area.

According to previous research, ISE from satellite images vary in different seasons due to the seasonal changes of vegetation. Impervious surfaces were reported to be overestimated in wintertime, when tree canopies are at their minimum (Weng et al. 2009). Moreover, different ISE methods are sensitive to this seasonal change in different ways. For instance, linear spectral mixture analysis is more sensitive to seasons, while the regression tree model is less sensitive to this change (Wu and Yuan 2007). Nevertheless, both the work done by Wu and Yuan (2007) and Weng et al. (2009) was conducted at the midlatitude region where the plant phenology undergoes dramatic changes in different seasons. In contrast, tropical and subtropical areas have a significantly different situation compared with temperate regions in most previous study sites. Therefore, it is still not clear whether the seasonal effects observed in previous studies can be compared with those in tropical and subtropical areas.

This study aims to address this question and assess the seasonal effects of ISE in tropical and subtropical regions. In this study, four cities including Guangzhou (China), Cape Town (South Africa), Mumbai (India), and Sao Paulo (Brazil), were selected for analyzing the seasonal effects of impervious surface estimation using Landsat TM/ETM+ images. Two widely used machine-learning methods, ANN and SVM, are applied to extract the impervious surfaces. ANN and SVM have been frequently reported to obtain superior performance over other classifiers in remote sensing studies, and thus are employed in this study. The confusion matrix is employed to assess the accuracy of ISE and thus to quantitatively investigate the seasonal effects on ISE of the climate zone in tropical and subtropical urban areas.

4.2 Datasets and Methodology

In tropical and subtropical areas, it is very difficult to choose an image without any clouds even during dry seasons (Fan et al. 2008). The images selected for this study are accompanied by small areas of clouds that are difficult to remove by image processing algorithms because they are very small and thin (Hagolle et al. 2010). During the selection of Landsat images for different seasons in the four study sites, it was difficult to find images without any contamination from clouds. Consequently, we tried to select the relatively best images for each season of each study site. We found that some images are free of clouds, while some are affected by a small amount of clouds and the maximum cloud coverage is 6% according to the United States Geological Survey (USGS) archive record. In order to investigate the seasonal effects of impervious surface estimation from satellite images, four scenes of Landsat TM/ETM+ images were chosen for four different seasons in this study. Table 4.1 shows that the seasonal satellite data is selected from different cyclical seasons and different dry and wet seasons. Additionally, note that the climate characteristics are different in the four cities. In Guangzhou, Mumbai, and Sao Paulo, hot seasons are generally wet, and cool seasons are dry. In Cape Town, hot seasons are dry, and cool seasons are wet.

The specification of Landsat ETM+ images can be found in Chapter 3, Section 3.2.1 of this book. Moreover, Landsat calibration was also applied using published postlaunch gains and offsets (Chander et al. 2007, 2009). The calibration was conducted using the software calibration utilities module of ENVI 4.7. We assumed that the atmospheric conditions were clear and

TABLE 4.1

Landsat Images for Four Study Sites

Imagery Date	Season		Imagery Date	Season	
	Guangzhou			Mumbai	
2009-01-10	Winter	Dry	2011-01-30	Winter	Dry
2009-05-02	Spring	Wet	2010-04-17	Spring	Dry
2009-08-22	Summer	Wet	2010-05-03	Summer	Wet
2009-10-09	Autumn	Dry	2010-10-26	Autumn	Wet
	Sao Paulo			Cape Town	
2010-02-05	Summer	Wet	2011-01-03	Summer	Dry
2010-04-18	Autumn	Dry	2011-04-09	Autumn	Wet
2010-08-24	Winter	Dry	2011-06-02	Winter	Wet
2010-11-28	Spring	Wet	2011-10-02	Spring	Dry

homogeneous and the small area of clouds would not significantly impact the whole scene of the image, and thus no atmospheric correction was performed (Wu and Murray 2003).

With careful examination of the four study cases, five land use types were defined to conduct the classification procedure according to the landscape of the study area. These land cover types include water, vegetation, bare soil, dark impervious surfaces, and bright impervious surfaces. ANN and SVM were then applied to classify the four seasonal data of ETM+ images. After seven land use classes were obtained, a combining operation was employed to reclassify the five land use types into two types: impervious surfaces and nonimpervious surfaces. During this stage, water body, vegetation cover, and bare soils were combined to form the nonimpervious surface class, while dark and bright impervious surfaces were combined to form the impervious surface class.

In order to obtain training and test samples, a visual interpretation procedure was used by visually comparing the four seasons of Landsat images and very-high-resolution images from Google Earth from the history data according to the Landsat images. However, terrains may appear differently in different seasons due to the seasonal changes of the plants, weather, and solar radiation, while the actual area and distribution of the impervious surfaces may be relatively stable, as all the images are selected in the same year. In this study, we assumed that the actual area and distribution of impervious surfaces stay the same during the year of the imagery dates. Therefore, different training sample sets are designed for different seasons of images, while only one testing dataset is sampled for all seasonal images. The samples included two main types: impervious surfaces and nonimpervious surfaces. First, to sample the training data for impervious surfaces, bright impervious surfaces and dark impervious surfaces were sampled separately.

Second, three types of land cover were sampled for nonimpervious surfaces: water body, vegetation, and bare soils. After the classification procedure, a postclassification is applied to combine all the land cover types into impervious surfaces and nonimpervious surfaces accordingly. For the testing data, both samples for the impervious surfaces and nonimpervious surfaces were obtained normally from each subtype of land cover. Lastly, the overall accuracy and Kappa coefficient based on the confusion matrix were employed to assess the accuracy of the impervious surface estimation (Jensen 2007).

Additionally, as one of the typical approaches, LSMA is also employed to compare with ANN and SVM. To apply LSMA, water surfaces in all the Landsat images of the four cities was first masked out with an unsupervised classification process (Wu and Murray 2003). Then, minimum noise fraction (MNF) transform was applied and the first three components were employed to analyze the feature space to select four endmembers (i.e., vegetation, bare soil, low albedo, and high albedo) from the images based on the VIS conceptual model (Ridd 1995). Next, using the SMA approach, four fraction images were calculated from each image. By adding the low albedo and high albedo fraction images, the fraction image of impervious surfaces could be obtained according to previous research (Wu and Murray 2003). However, in order to compare with the impervious surface results from ANN and SVM, the sub-pixel estimation of impervious surfaces from LSMA was transformed into a per-pixel level by the following rule: pixels with a portion of impervious surfaces equal to or higher than 50% were treated as impervious surface pixels, while all others were treated as nonimpervious surface pixels. The per-pixel level of impervious surfaces derived from LSMA could then be assessed by the same test sample sets that were applied to the results from ANN and SVM for comparison.

4.3 Results and Discussion

4.3.1 Guangzhou

Figure 4.1 shows the estimate of impervious surfaces with ANN and Table 4.2 shows the accuracy assessment result of the extracted impervious surfaces using the ANN method, which shows a clearer illustration of the patterns. In contrast with the previously reported research in the midlatitude region (Weng et al. 2009; Wu and Yuan 2007), the winter imagery (January 10) obtained the best accuracy, with an overall accuracy of 91.44% and a Kappa coefficient of 0.8287. In spring, the accuracy was also good (overall accuracy, 88.25%; Kappa coefficient, 0.7651). The lowest accuracy happened in the autumn period (October 9), with an overall accuracy of only 83.30% and a Kappa coefficient of 0.6615. Summertime (August 22), which was reported

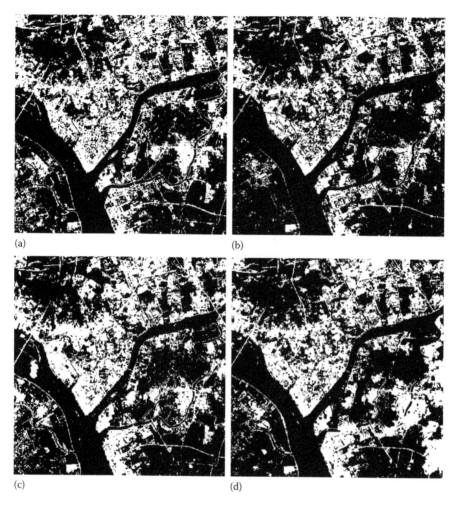

FIGURE 4.1
Estimation of impervious surfaces in Guangzhou using ANN. Bright areas show impervious surfaces and dark areas show nonimpervious surfaces. (a) 2009-01-10, (b) 2009-05-02, (c) 2009-08-22, and (d) 2009-10-09.

to be the best season for ISE in temperate regions, only had 84.50% overall accuracy and a Kappa coefficient of 0.6873. Generally, dry seasons obtained much higher accuracy than wet seasons.

Figure 4.2 shows the results with SVM and Table 4.3 compares the accuracy assessment of ISE from the SVM approach. The highest accuracy occurred in the wintertime, when the overall accuracy was 92.0028% and the Kappa coefficient was 0.8394. Both the spring and autumn generated low accuracy in this experiment, with the overall accuracy only about 83% and a Kappa coefficient of about 0.66. The summer image produced a slightly higher accuracy

TABLE 4.2

Accuracy Assessment of the Classification Results by ANN

Image Date	Season		OA (%)	Kappa
2009-01-10	Winter	Dry	91.4367	0.8287
2009-05-02	Spring	Wet	88.2519	0.7651
2009-08-22	Summer	Wet	84.5011	0.6873
2009-10-09	Autumn	Wet	83.2979	0.6615

Note: Kappa = Kappa coefficient; OA = overall accuracy.

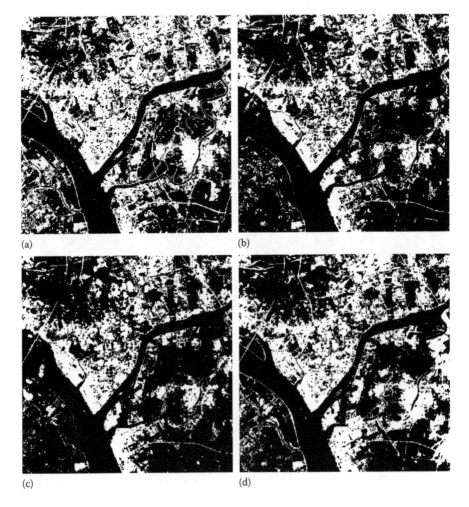

FIGURE 4.2
Estimation of impervious surfaces in Guangzhou using SVM. Bright areas show impervious surfaces and dark areas show nonimpervious surfaces. (a) 2009-01-10, (b) 2009-05-02, (c) 2009-08-22, and (d) 2009-10-09.

TABLE 4.3

Accuracy Assessment of the Classification Results by SVM

Image Date	Season		OA (%)	Kappa
2009-01-10	Winter	Dry	92.0028	0.8394
2009-05-02	Spring	Wet	83.0149	0.6588
2009-08-22	Summer	Wet	84.9965	0.7000
2009-10-09	Autumn	Wet	83.2272	0.6592

Note: Kappa = Kappa coefficient; OA = overall accuracy.

than those of the spring and autumn, but the accuracy is still much lower than that of the wintertime. The SVM results illustrate the same consistent relationship between climate zone and ISE accuracy as in the ANN results.

However, LSMA produced a rather different pattern of impervious surface results, as shown in Figure 4.3 and Table 4.4, where the results are better in wet seasons than in dry seasons. Moreover, the accuracy is generally lower than that from ANN and SVM in all four seasons. By interpreting the results shown in Figure 4.3, a noticeable underestimation in winter and summer, as well as a noticeable overestimation in spring and autumn, can be observed. Actually, this error mainly comes from the threshold set for transforming the impervious surface results from the subpixel level to the per-pixel level. This study used an empirical threshold of 50%, which remains an unaddressed issue. In order to assess the accuracy of impervious surfaces derived from LSMA at a per-pixel level, this important threshold needs to be further investigated.

4.3.2 Mumbai

The ISE results and accuracy assessment using ANN in Mumbai are presented in Figure 4.4 and Table 4.5, and show that the best result comes from spring (April 17) with an overall accuracy of 91.18% and a Kappa value of 0.8219. The spring result is slightly better than that in winter. Results in winter (January 30) and spring (April 17) have a noticeably higher accuracy than that from summer (May 3) and autumn (October 26). Autumn results have the lowest accuracy because a number of impervious surface pixels were misclassified as nonimpervious surface pixels and there is an observable underestimation in Figure 4.4d. Note that misclassifications in rainy seasons (summer and autumn) happen more frequently, which can be observed in the figures. Nevertheless, there is an overall overestimation in all four season results. This is caused by the complex land covers pattern in Mumbai. In this study area, residential and commercial areas are very small and fragmented, while greening areas are also fragmentally located among these impervious surfaces. Therefore, there are numerous mixed pixels in the Landsat images at 30 m × 30 m resolution. With a per-pixel method, some mixed pixels are treated as impervious surface pixels. As a result, overestimation occurs even

(a)   (b)

(c)   (d)

FIGURE 4.3
Estimation of impervious surfaces in Guangzhou using LSMA. Bright areas show impervious surfaces and dark areas show nonimpervious surfaces. (a) 2009-01-10, (b) 2009-05-02, (c) 2009-08-22, and (d) 2009-10-09.

TABLE 4.4

Accuracy Assessment of the Classification Results by LSMA

Image Date	Season		OA (%)	Kappa
2009-01-10	Winter	Dry	61.57	0.2208
2009-05-02	Spring	Wet	84.78	0.6906
2009-08-22	Summer	Wet	61.01	0.2184
2009-10-09	Autumn	Wet	78.98	0.5674

Note: Kappa = Kappa coefficient; OA = overall accuracy.

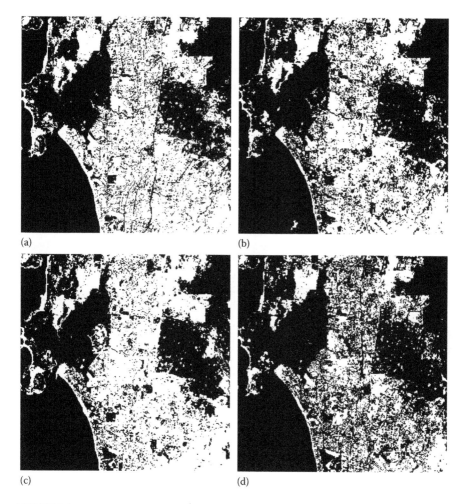

FIGURE 4.4
Estimation of impervious surfaces in Mumbai using ANN. Bright areas show impervious surfaces and dark areas show nonimpervious surfaces. (a) 2011-01-30, (b) 2010-04-17, (c) 2010-05-03, and (d) 2010-10-26.

TABLE 4.5

Accuracy Assessment of the Classification Results in Mumbai by ANN

Date	Season		OA (%)	Kappa
2010-01-30	Winter	Dry	90.78	0.8154
2010-04-17	Spring	Dry	91.18	0.8219
2010-05-03	Summer	Wet	88.59	0.7708
2010-10-26	Autumn	Wet	85.84	0.7114

Note: Kappa = Kappa coefficient; OA = overall accuracy.

in the dry seasons. However, the comparison between different seasons is reliable since this phenomenon exists in all the seasons and can be ignored during the comparison.

In the results from SVM, there are some slight differences, which are shown in Figure 4.5 and Table 4.6. The best result comes from the winter image, with an overall accuracy of 92.23% and a Kappa coefficient of 0.8450. The lowest result is from the autumn image, with an overall accuracy of 86.25% and a Kappa coefficient of 0.7201. Generally, dry seasons (winter and spring) are more suitable for the mapping of impervious surfaces than wet seasons (summer and autumn). However, the overestimation phenomenon still exists

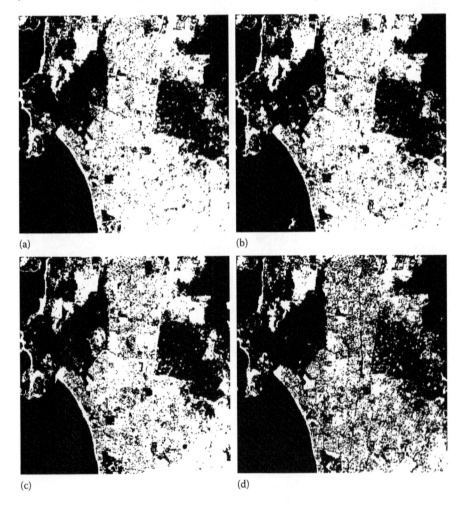

(a)   (b)

(c)   (d)

FIGURE 4.5
Estimation of impervious surfaces in Mumbai using SVM. Bright areas show impervious surfaces and dark areas show nonimpervious surfaces. (a) 2011-01-30, (b) 2010-04-17, (c) 2010-05-03, and (d) 2010-10-26.

TABLE 4.6

Accuracy Assessment of the Classification Results in
Mumbai by SVM

Date	Season		OA (%)	Kappa
2010-01-30	Winter	Dry	92.23	0.8450
2010-04-17	Spring	Dry	91.83	0.8358
2010-05-03	Summer	Wet	87.14	0.7421
2010-10-26	Autumn	Wet	86.25	0.7201

Note: Kappa = Kappa coefficient; OA = overall accuracy.

in all four seasons due to the mixed pixels. Considering all the results, both ANN and SVM get a generally consistent result for ISE. Winter is the most suitable season for ISE using the per-pixel classification approach, while summer and autumn are not appropriate for ISE due to high precipitation.

Comparatively, the results from LSMA are less accurate than those from ANN and SVM, as shown in Figure 4.6 and Table 4.7. Nevertheless, the overall pattern is generally consistent with other results that dry seasons had a higher accuracy than wet seasons. The highest accuracy came from the winter image, with an overall accuracy of 86.33% and a Kappa value of 0.7237, while the lowest accuracy came from the summer image, with an overall accuracy of 55.50% and a Kappa value of only 0.1029. As shown in Figure 4.6, an overall underestimation can be observed from all four seasons, which indicates that the empirical threshold used to transform the LSMA results from the subpixel to per-pixel level was possibly too high and should be decreased.

4.3.3 Sao Paulo

With the ANN method, the best ISE result in Sao Paulo was obtained from the winter image (August 24) with an overall accuracy of 94.14% and a Kappa value of 0.8829 (Table 4.8). Spring (February 5) had the lowest accuracy (overall accuracy: 87.21; Kappa: 0.7453) because more impervious surface pixels were misclassified as nonimpervious surfaces pixels. This can also be observed in Figure 4.7a, where there is an underestimation on the northwest part of the study area. ISE results in Sao Paulo are generally higher than those in other cities. However, shadows from tall buildings in this area produced more misclassification, which can be observed in the upper-middle part of the study area in Figure 4.7. The major spectral confusions are those between bare soil and both dark and bright impervious surfaces and between water surfaces and shadows from tall buildings. Results in dry seasons are generally better than that in wet seasons. However, the autumn (dry) accuracy is slightly lower than the spring (wet) accuracy as observed, which was derived from the underestimation of impervious surfaces in the autumn image.

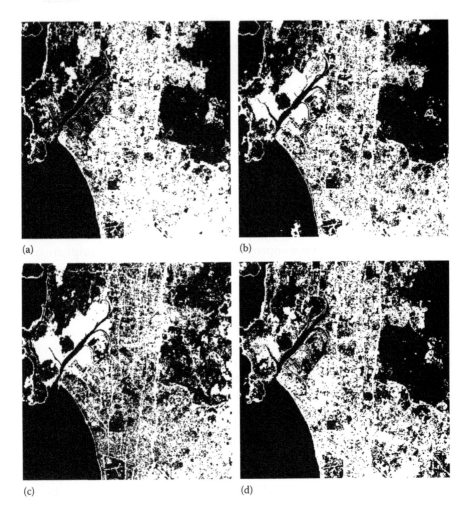

(a)   (b)

(c)   (d)

FIGURE 4.6
Estimation of impervious surfaces in Mumbai using LSMA. Bright areas show impervious surfaces and dark areas show nonimpervious surfaces. (a) 2011-01-30, (b) 2010-04-17, (c) 2010-05-03, and (d) 2010-10-26.

TABLE 4.7

Accuracy Assessment of the Classification Results in Mumbai by LSMA

Date	Season		OA (%)	Kappa
2010-01-30	Winter	Dry	86.33	0.7237
2010-04-17	Spring	Dry	69.74	0.3836
2010-05-03	Summer	Wet	55.50	0.1029
2010-10-26	Autumn	Wet	73.38	0.4502

Note: Kappa = Kappa coefficient; OA = overall accuracy.

TABLE 4.8

Accuracy Assessment of the Classification Results in Sao Paulo by ANN

Date	Season		OA (%)	Kappa
2010-02-05	Summer	Wet	87.21	0.7453
2010-04-18	Autumn	Dry	92.63	0.8529
2010-08-24	Winter	Dry	94.14	0.8829
2010-11-28	Spring	Wet	93.43	0.8688

Note: Kappa = Kappa coefficient; OA = overall accuracy.

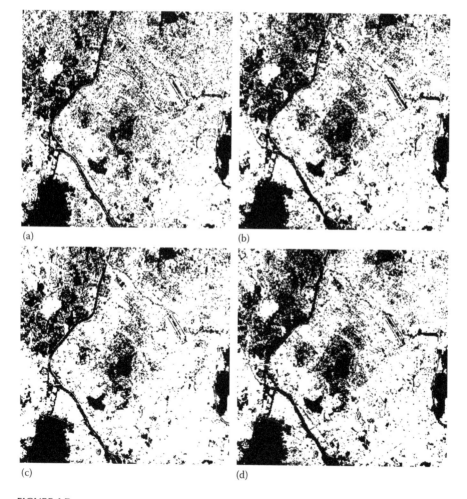

(a)

(b)

(c)

(d)

FIGURE 4.7

Estimation of impervious surfaces in Sao Paulo using ANN. Bright areas show impervious surfaces and dark areas show nonimpervious surfaces. (a) 2010-02-05, (b) 2010-04-18, (c) 2010-08-24, and (d) 2010-11-28.

The results using SVM have a consistent pattern with those using ANN in Sao Paulo (Figure 4.8 and Table 4.9). The winter image had the highest accuracy, with an overall accuracy of 94.14 and a Kappa value of 0.8829, while the lowest result was again from spring, with an overall accuracy of 88.99% and a Kappa value of 0.7805. The results demonstrate that different methods did not have significant impact on the ISE using seasonal images, and the overall underestimation appears in the results from both ANN and SVM.

When using LSMA, an inverse pattern was observed as in the Sao Paulo case (Figure 4.9 and Table 4.10), where wet seasons (summer and spring) obtained higher accuracy than dry seasons (autumn and winter). The highest

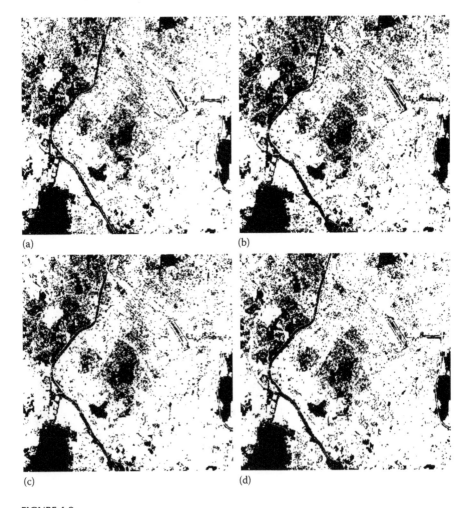

(a)

(b)

(c)

(d)

FIGURE 4.8

Estimation of impervious surfaces in Sao Paulo using SVM. Bright areas show impervious surfaces and dark areas show nonimpervious surfaces. (a) 2010-02-05, (b) 2010-04-18, (c) 2010-08-24, and (d) 2010-11-28.

TABLE 4.9

Accuracy Assessment of the Classification Results in
Sao Paulo by SVM

Date	Season		OA (%)	Kappa
2010-02-05	Summer	Wet	88.99	0.7805
2010-04-18	Autumn	Dry	91.56	0.8315
2010-08-24	Winter	Dry	94.14	0.8829
2010-11-28	Spring	Wet	92.18	0.8440

Note: Kappa = Kappa coefficient; OA = overall accuracy.

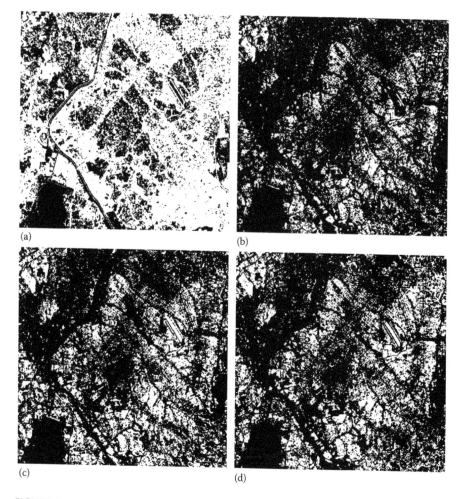

(a) (b) (c) (d)

FIGURE 4.9

Estimation of impervious surfaces in Sao Paulo using LSMA. Bright areas show impervious
surfaces and dark areas show nonimpervious surfaces. (a) 2010-02-05, (b) 2010-04-18, (c) 2010-
08-24, and (d) 2010-11-28.

TABLE 4.10

Accuracy Assessment of the Classification Results in Sao Paulo by LSMA

Date	Season		OA (%)	Kappa
2010-02-05	Summer	Wet	76.15	0.5234
2010-04-18	Autumn	Dry	65.97	0.3206
2010-08-24	Winter	Dry	73.32	0.4671
2010-11-28	Spring	Wet	75.67	0.5138

Note: Kappa = Kappa coefficient; OA = overall accuracy.

accuracy came from the summer image, with an overall accuracy of 76.15% and a Kappa coefficient of 0.5234. The lowest accuracy was derived from the autumn image, with an overall accuracy of 65.97% and a Kappa coefficient of 0.3206. Underestimation was also observed generally for all four seasons, indicating the high threshold value for the transformation of subpixel impervious surfaces from LSMA.

4.3.4 Cape Town

The seasonal effect of ISE in Cape Town is more complex due to the Mediterranean climate. In autumn and winter when the average temperature is low, the weather is wet, while spring and summer are dry seasons and the temperature is high. Therefore, the seasonal changes of plants are not consistent with those in other tropical and subtropical areas. These climatology and phenology characteristics may have impacts on ISE in this region. Similarly, impervious surfaces were estimated comparatively using two methods, ANN and SVM, to investigate the impacts on ISE. The ISE results of Cape Town using ANN are shown in Figure 4.10 and the related accuracy assessment is shown in Table 4.11. Generally, in summer and autumn, more nonimpervious surface pixels were misclassified as impervious surface pixels and thus there was an overestimation in these two images and the accuracy is relatively lower. The lowest accuracy comes from summer, with an overall accuracy of 88.83% and a Kappa coefficient of 0.7746. Winter and spring obtained generally a higher accuracy by reducing the misclassification of nonimpervious surface pixels, and the best result was from the spring image, with an overall accuracy of 93.24% and a Kappa coefficient of 0.8634.

The ISE results in Cape Town using SVM are shown in Figure 4.11 and the accuracy assessment is shown in Table 4.12. The results are generally consistent with the results from ANN with only a slight difference in the winter and spring results. Overall overestimation was also observed in summer and autumn produced by the incorrect classification of nonimpervious surface pixels as impervious surface pixels. The lowest accuracy was obtained from the summer image, with an overall accuracy of 88.92% and a Kappa coefficient of 0.7758. Winter and spring also had generally better

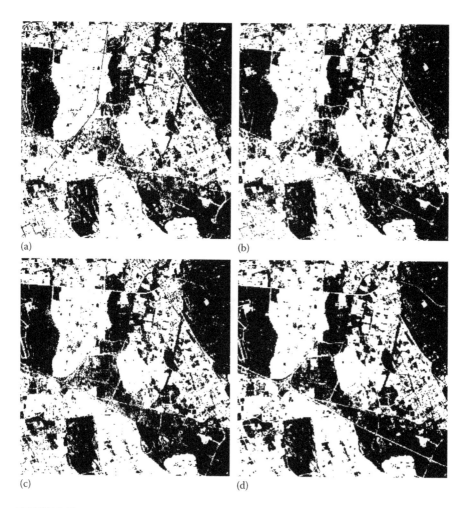

FIGURE 4.10
Estimation of impervious surfaces in Cape Town using ANN. Bright areas show impervious surfaces and dark areas show nonimpervious surfaces. (a) 2011-01-03, (b) 2011-04-09, (c) 2011-06-02, and (d) 2011-10-02.

TABLE 4.11

Accuracy Assessment of the Classification Results in Cape Town by ANN

Date	Season		OA (%)	Kappa
2010-01-03	Summer	Dry	88.83	0.7746
2010-04-09	Autumn	Wet	89.58	0.7906
2010-06-02	Winter	Wet	92.11	0.8405
2010-10-02	Spring	Dry	93.24	0.8634

Note: Kappa = Kappa coefficient; OA = overall accuracy.

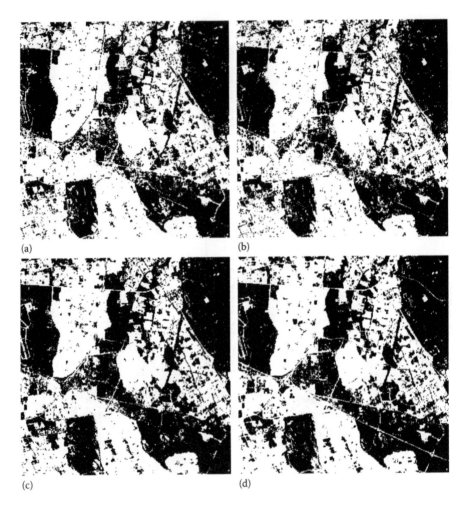

FIGURE 4.11
Estimation of impervious surfaces in Cape Town using SVM. Bright areas show impervious surfaces and dark areas show nonimpervious surfaces. (a) 2011-01-03, (b) 2011-04-09, (c) 2011-06-02, and (d) 2011-10-02.

TABLE 4.12

Accuracy Assessment of the Classification Results in Cape Town by ANN

Date	Season		OA (%)	Kappa
2010-01-03	Summer	Dry	88.92	0.7758
2010-04-09	Autumn	Wet	89.59	0.7902
2010-06-02	Winter	Wet	92.68	0.8514
2010-10-02	Spring	Dry	92.49	0.8482

Note: Kappa = Kappa coefficient; OA = overall accuracy.

results than summer and autumn, while the best result did not come from spring but from the winter season, with an overall accuracy of 92.68% and a Kappa coefficient of 0.8514. The results indicate that different estimation methods may have some influence on the ISE results, but this influence is not significant compared with the impacts of seasonal changes of climatology and phenology.

Nevertheless, the results derived from LSMA in Cape Town seem unreliable due to the rather low accuracy (Figure 4.12 and Table 4.13). The highest accuracy was from the spring image, with an overall accuracy of 55.87% and a Kappa value of 0.139, which are too low to be compared with the results

(a)   (b)   (c)   (d)

FIGURE 4.12
Estimation of impervious surfaces in Cape Town using LSMA. Bright areas show impervious surfaces and dark areas show nonimpervious surfaces. (a) 2011-01-03, (b) 2011-04-09, (c) 2011-06-02, and (d) 2011-10-02.

TABLE 4.13

Accuracy Assessment of the Classification Results in Cape Town by LSMA

Date	Season		OA (%)	Kappa
2010-01-03	Summer	Dry	54.55	0.1115
2010-04-09	Autumn	Wet	52.02	0.0138
2010-06-02	Winter	Wet	51.36	0.013
2010-10-02	Spring	Dry	55.87	0.139

Note: Kappa = Kappa coefficient; OA = overall accuracy.

from ANN and SVM. Therefore, there should be some critical error sources during the application of LSMA, such as the endmember selection and the threshold determination of transforming subpixel estimation to per-pixel estimation.

4.4 Discussion

Plant phenology was considered to be the main cause of the seasonal changes of ISE, and ISE in leaf-on seasons are more accurate than those estimated in leaf-off seasons (Weng 2012; Weng et al. 2009; Wu and Yuan 2007). As reported previously, in summer when plant leaves are on the trees, most areas of bare soils are overlaid by trees, grass, and crops. Hence, the confusion between bare soils and bright impervious surfaces is reduced greatly, and thus the estimated impervious surfaces are more accurate. However, the case in a tropical and subtropical region turns out to be much more complicated. First, seasonal plant changes occur much less frequently in tropical and subtropical areas, where most of the plants are evergreen throughout the whole year, so there is no boundary between the so-call leaf-on and leaf-off seasons. However, the plants may change in flowering seasons depending on the temperature and precipitation. Second, water bodies become seasonally changeable in rainy and dry seasons in tropical and subtropical areas but the specific characteristics of these seasonal changes may be also different in different climate zones due to the distribution of wet seasons and dry seasons. In this study, taking Guangzhou as an example, we try to discuss the seasonal change characteristic of some typical land covers.

The city of Guangzhou has a humid subtropical climate. Due to the dramatic urbanization and irregular urban planning and management, there are many VSAs in this region. In wet seasons, rainwater fills many areas of different sizes that will be dry in dry seasons. These areas are the so-called VSAs identified by a hydrological scientist (Frankenberger et al. 1999).

Moreover, the water level in lakes and rivers in dry seasons will be much lower than that in rainy seasons. Thus, theoretically, in dry seasons, more bare soils are exposed and are likely to be confused with bright impervious surfaces. Third, cloud coverage is another important factor. Even in dry seasons, small pieces of clouds appear from time to time. Thus, optical satellite images that are cloud-free are almost not available throughout the year. These clouds become targets that are confused with bright impervious surfaces.

In order to investigate the seasonal effects of impervious surfaces, about 60 pixels of each typical land cover types are carefully selected and their average spectral characteristics are shown in Figure 4.13, depicting a typical situation of spectral reflectance in a subtropical monsoon urban city. It is impressive that there are two groups of spectral confusion. One group includes the clouds and bright impervious surfaces. Both of these two targets have high reflectance in the six bands. The other group has water bodies shade, and dark impervious surfaces. The patterns of these three targets are very similar, with low reflectance in the six spectral bands. This is quite an important factor to ISE, because water bodies increase in rainy seasons with more VSA. As a result, the confusion between water and dark impervious surfaces increases in rainy seasons. Another impressive finding is that the average reflectance characteristics of bare soils seem to be much lower than shown in previous reports (Weng et al. 2009; Wu and Yuan 2007). This is not surprising, since bare soils in a humid area are almost wet soil because of the humid climate. Therefore, it is much easier to discriminate the bare soil from either bright or dark impervious surfaces.

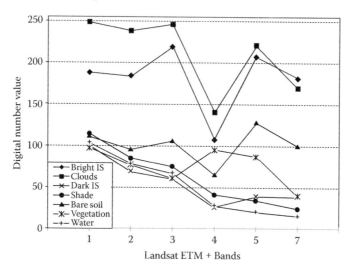

FIGURE 4.13
Average digital numbers of different land use types in humid areas.

To illustrate the characteristics of the seasonal changes of typical targets in the subtropical monsoon region discussed above, four locations in the study case of this chapter are selected, representing different types of terrain. Images of the four dates of the four blocks, together with the related estimated impervious surfaces using the ANN method, are correspondingly shown in Figure 4.14. First, the vegetation subfigure shows that the vegetated area keeps relatively stable in different seasons and vegetated areas have little impact on the estimated impervious surfaces. Second, the bare soil subfigure reveals that even though bare soils undergo some changes in different seasons, they can be easily recognized as nonimpervious surfaces; however, water bodies are mistakenly classified as dark impervious surfaces. Third, the VSA subfigure demonstrates a similar case in the bare soil subfigure, where the water in VSA in May and August is wrongly treated as dark impervious surfaces. Fourth, the cloud and shadow subfigure demonstrates the confusion between clouds and bright impervious surfaces and the confusion between shade and dark impervious surfaces. Clouds and related shade in August are misclassified as impervious surfaces.

Generally, in dry seasons, VSAs are empty of water and clouds are much less present, and therefore, less water is confused with dark impervious surfaces and fewer clouds are confused with bright impervious surfaces. Nevertheless, in rainy seasons, all VSA regions are filled with water, and many more clouds occur, resulting in a number of cloud shades. In this study in particular, the autumn image was taken at the beginning of the dry season (October 9) when most water in VSA regions still remains, and therefore it was treated as the wet/rainy season. Thus, the spectral confusion between water/shades and dark impervious surfaces, and between clouds and bright impervious surfaces, increase greatly. As a result, the dry season is more appropriate for the estimation of impervious surfaces in subtropical monsoon areas.

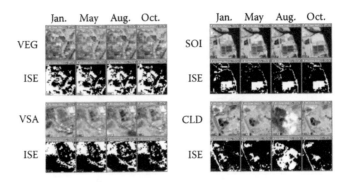

FIGURE 4.14
Effects of seasonal changes of typical targets, VEG = vegetation, SOI = bare soil, VSA = variable source area, CLD = clouds and shade, ISE = impervious surface estimation.

4.5 Conclusion

Accurate ISE remains challenging due to the diversity of impervious surfaces, spectral confusion among various land covers, and the seasonal changes of vegetation and climate. Among the challenges, seasonal effects from the climate zones is one of the key issues that influence the accurate estimation of impervious surfaces. In this study, four scenes of Landsat TM/ETM+ images were carefully chosen for four different seasons in four typical cities—Guangzhou, Mumbai, Sao Paulo, and Cape Town—from tropical and subtropical areas, and two classification methods, the ANN and the SVM, were employed to extract the impervious surfaces from the images at the pixel level. The commonly used LSMA was also employed to estimate impervious surfaces at a subpixel level with an additional transformation to per-pixel level estimation. The experimental results demonstrate quite a unique view of seasonal effects in tropical and subtropical areas that is different from that in midlatitude or temperate areas according to previous research (Weng et al. 2009; Wu and Yuan 2007). According to the results, in tropical and subtropical regions, winter and spring are generally the better seasons to estimate impervious surfaces from optical remote sensing images compared with summer and autumn. Winter and spring are generally the dry seasons in tropical and subtropical regions and the temperature is relatively lower. With a specific investigation in Guangzhou, we found that in winter there is a small amount of cloud and most of the VSAs are not filled with water. Even though more bare soils in the VSA are exposed, they can be easily identified because most are actually not dry soils as in the midlatitude areas. Therefore, satellite images are the most appropriate for estimating impervious surfaces. On the other hand, autumn images had the lowest accuracy of impervious surfaces due to the cloud coverage and water in VSAs. Autumn is a rainy season in a subtropical monsoon region, in which clouds occur very often and VSAs are always filled with water. Consequently, clouds are confused with bright impervious surfaces due to their similar high reflectance, and more water in VSAs is confused with dark impervious surfaces due to their similar low reflectance.

Moreover, the seasonal sensitivity of the two methods is compared. Both ANN and SVM methods show a general consistency in accurately depicting seasonal changes. However, both methods indicated that wintertime is the best season for ISE with satellite images in subtropical monsoon regions. The limitations of this study come from the methodology, which is generally based on a per-pixel level. In urban and suburban areas, one pixel with a size of 30 m × 30 m does not necessarily include only impervious materials or nonimpervious materials (Weng 2012; Wu and Murray 2003). In this case, the use of per-pixel methods would obtain a result with lower accuracy.

5

Assessing the Urban Land Cover Complexity

5.1 Introduction

Urban land use/land cover (LULC) classification is important for monitoring urbanization and its impacts on the environment (Lu and Weng 2006). However, the accuracy needs to be improved and it is still a challenge due to the diversity of LULC types (Lu et al. 2010; Lu and Weng 2006). Various satellite data has been applied to classify LULC using coarse, medium, and high-resolution images (Myint et al. 2011). However, there are some open problems that need to be addressed in order to improve the classification accuracy (Lu et al. 2004; Lu and Weng 2006). First, bare soils or sands were reported to be often confused with bright impervious surfaces (e.g., cool roofs and new concrete roads), while shade and water were often confused with dark impervious surfaces (e.g., asphalt and old concrete roads). These confusions are caused by the similar spectral reflectance of different materials (Lu and Weng 2006). Second, clouds and their shadows are considered a difficult issue to deal with in subtropical humid regions, where cloudy and rainy weather occurs throughout the entire year. Both these problems lower the accuracy of the LULC classification in subtropical humid urban areas. To deal with these problems, SAR remote sensing data is employed and combined with optical images to provide complementary information to help differentiate similar spectral reflectance of different LULC types and help identify the LULC information in cloudy areas. However, before combining the optical and SAR remote sensing data, the spectral confusion between various land covers should be investigated to better understand the situation in tropical and subtropical urban areas. This chapter aims to investigate the urban land cover complexity using optical remote sensing data to determine the confusion between different land covers in four tropical and subtropical cities.

5.2 Datasets and Methodology

Since the optical images will be combined with the SAR images to estimate impervious surfaces in the following chapters, the selection of optical images for this chapter was conducted according to the availability of SAR images in the corresponding study area. Similar to the study in Chapter 4, five land use types were defined to conduct the classification procedure according to the landscape of the study area. These land cover types include water (WAT), vegetation (VEG), bare soil (SOI), dark impervious surfaces (DIS), and bright impervious surfaces (BIS). Moreover, in order to analyze the impacts of classification methods to better understand the real spectral confusions in tropical and subtropical areas, two popular machine-learning methods, ANN and SVM, were employed to conduct the LULC classification. The training and test samples were the same sets used in Chapter 4. The confusion matrix was used as the main tool to investigate the spectral confusion between various land cover types. Additionally, we assume that the collected training and test samples had some unavoidable errors, and in order to eliminate the influence of these errors, the analysis and discussion will focus on the misclassifications of more than 10 pixels between two land cover classes.

5.3 Results and Discussion

5.3.1 Guangzhou

In the Guangzhou study area, the LULC classification results using ANN and SVM are given in Figure 5.1, showing a generally consistent result of urban land covers. The noticeable difference between the two results is the

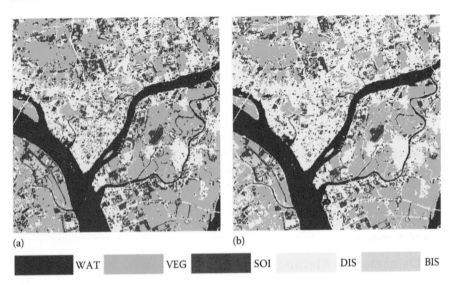

(a)   (b)

| WAT | VEG | SOI | DIS | BIS |

FIGURE 5.1
LULC classification in Guangzhou: (a) ANN and (b) SVM.

total number of bare soil pixels and their distribution. It can be observed that more bare soil pixels over the whole area were classified using ANN. These bare soil pixels were classified as vegetation and dark impervious surface pixels in the SVM result. Moreover, most of these inconsistent pixels were located on or near the boundaries between different land covers. In general, SVM produced a better LULC classification result with less impact by the spectral confusion between various land covers.

In order to investigate the specific confusion between various land covers, the confusion matrix using the testing samples were calculated for both the ANN and SVM methods (Tables 5.1 and 5.2). From the confusion matrix of the ANN result, three major pairs of land covers were identified as being more easily confused. First, vegetation and bare soil tended to be confused with each other, with 28 pixels of vegetation misclassified as bare soil pixels and 12 pixels of bare soil misclassified as vegetation pixels. Second, bare soil seemed more easily confused with dark impervious surfaces, with 28 soil pixels misclassified as dark impervious surface pixels and 14 dark impervious surface pixels misclassified as bare soil pixels. Third, bright impervious surface pixels were also confused with bare soil, with 26 pixels of bright impervious surface misclassified as bare soil pixels. Generally, the overall

TABLE 5.1

Confusion Matrix of ISE in Guangzhou Using ANN

	WAT	VEG	SOI	DIS	BIS
WAT	102	5	0	0	0
VEG	0	90	12	1	3
SOI	0	28	122	14	26
DIS	0	1	28	153	2
BIS	0	0	0	0	138

Note: Overall accuracy: 83.45%, Kappa coefficient: 0.7911. WAT, water; VEG, vegetation; SOI, bare soil; DIS, dark impervious surfaces; BIS, bright impervious surfaces.

TABLE 5.2

Confusion Matrix of ISE in Guangzhou Using SVM

	WAT	VEG	SOI	DIS	BIS
WAT	102	3	0	0	0
VEG	0	98	1	0	3
SOI	0	17	125	11	27
DIS	0	6	36	157	2
BIS	0	0	0	0	137

Note: Overall accuracy: 85.38%, Kappa coefficient: 0.8153.

accuracy of the ANN classification was 83.45% and the Kappa coefficient was 0.7911. The most accurately classified land cover was water surface. When using the SVM approach, these spectral confusions were significantly reduced (Table 5.2). For instance, only 17 pixels of vegetation were misclassified as bare soil pixels and very few bare soil pixels were misclassified as vegetation. In addition, only 11 pixels of dark impervious surface were misclassified as bare soil pixels. However, there was a slight increase of the number of misclassified pixels from bare soil to dark impervious surface and from bright impervious surface to base soil. Generally, the accuracy of SVM classification was higher than that of the ANN result, with an overall accuracy of 85.38% and a Kappa coefficient of 0.8153.

5.3.2 Mumbai

Two methods, ANN and SVM, were used to conduct the LULC classification, with the classification results shown in Figure 5.2. It can be observed that many more bare soil pixels were identified by the SVM, which was incorrect due the spectral confusion between bare soil and vegetation and between bare soil and impervious surfaces. Comparatively, ANN obtained a better result with noticeably reduced misclassifications of soil pixels.

The confusion matrix was calculated with the test samples, which is shown in Table 5.3. From the confusion matrix, several spectral confusions can be noticed. In the ANN result, vegetation tended to be misclassified as water surfaces (13 pixels) and dark impervious surfaces (32 pixels). Bare soil was easily misclassified as water surfaces (34 pixels), dark impervious surfaces

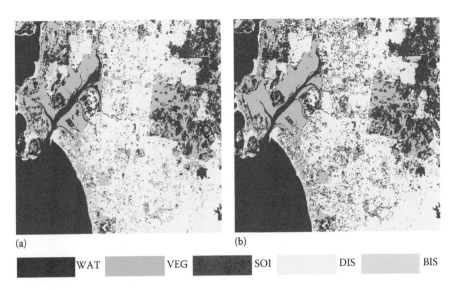

(a)   (b)

| WAT | VEG | SOI | DIS | BIS |

FIGURE 5.2
LULC classification in Mumbai: (a) ANN and (b) SVM.

TABLE 5.3

Confusion Matrix of ISE in Mumbai
Using ANN

	WAT	VEG	SOI	DIS	BIS
WAT	214	13	34	0	0
VEG	0	226	2	34	16
SOI	0	9	92	15	6
DIS	0	32	18	289	116
BIS	0	1	20	1	99

Note: Overall accuracy: 74.37%, Kappa coefficient: 0.6713.

(18 pixels), and bright impervious surfaces (20 pixels). Moreover, dark impervious surfaces and bright impervious surfaces were also confused with each other. Note that there was great confusion between dark and bright impervious surfaces, with 116 confused pixels of bright impervious surfaces misclassified as dark impervious surfaces. The overall accuracy was 74.37% and the Kappa coefficient was 0.6713. Meanwhile, in the SVM results (Table 5.4), vegetation was easily confused with bare soil (33 pixels) and dark impervious surfaces (28 pixels). Bare soil was also misclassified as dark impervious surfaces (47 pixels) or bright impervious surfaces (15 pixels). As well, there were incorrect classifications between dark and bright impervious surfaces: 117 pixels of bright impervious surfaces were misclassified as dark impervious surfaces. The overall accuracy was 74.21% and the Kappa coefficient was 0.6698. Generally, water surfaces were easy to identify in Mumbai, with less spectral confusion with other land cover types. Spectral confusion between impervious surfaces and nonimpervious surfaces was mainly due to vegetation and bare soil. Although there were a number of misclassified pixels between dark and bright impervious surfaces, they are subtypes of impervious surfaces, which had little impact on the mapping of impervious surfaces.

TABLE 5.4

Confusion Matrix of ISE in Mumbai
Using SVM

	WAT	VEG	SOI	DIS	BIS
WAT	211	0	4	0	0
VEG	3	219	0	16	3
SOI	0	33	100	31	19
DIS	0	28	47	290	117
BIS	0	1	15	2	98

Note: Overall accuracy: 74.21%, Kappa coefficient: 0.6698.

5.3.3 Sao Paulo

In this study case, land covers in Sao Paulo were less confused with each other. The noticeable confusions were between dark impervious surfaces and vegetation and bare soil, which influenced the LULC classification using both ANN and SVM. The results in Figure 5.3 demonstrate that a number of dark impervious surface pixels were misclassified as bare soil pixels, and thus produced an overall underestimation in the impervious surface results (Figure 5.3a). In contrast, SVM produced a more accurate LULC classification result, with much less misclassification of the bare soil. In fact, there were only a small number of bare soil pixels for such a highly developed urban area in this selected site, which is shown in the confusion matrix in Tables 5.5 and 5.6. Nevertheless, due to the spectral confusion between bare soil and

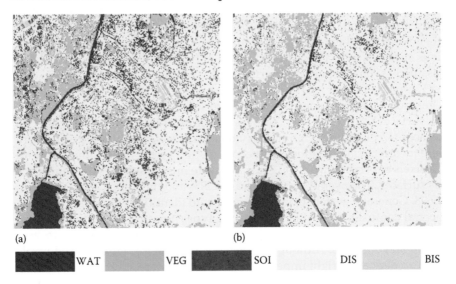

(a)   (b)

WAT   VEG   SOI   DIS   BIS

FIGURE 5.3
LULC classification in Sao Paulo: (a) ANN and (b) SVM.

TABLE 5.5

Confusion Matrix of ISE in Sao Paulo
Using ANN

	WAT	VEG	SOI	DIS	BIS
WAT	215	4	0	3	0
VEG	0	311	7	47	13
SOI	0	0	0	68	0
DIS	0	6	6	243	4
BIS	0	0	1	4	194

Note: Overall accuracy: 85.52%, Kappa coefficient: 0.8091.

TABLE 5.6

Confusion Matrix of ISE in Sao Paulo
Using SVM

	WAT	VEG	SOI	DIS	BIS
WAT	215	0	0	0	0
VEG	0	313	6	51	18
SOI	0	0	0	35	4
DIS	0	8	8	278	13
BIS	0	0	0	1	176

Note: Overall accuracy: 87.21%, Kappa coefficient: 0.8288.

dark impervious surfaces, a number of pixels were incorrectly identified as bare soil pixels.

From the confusion matrix of the ANN result, dark impervious surfaces were misclassified as vegetation (47 pixels) and bare soil (68 pixels), while bright impervious surfaces were misclassified as vegetation (13 pixels). Nevertheless, one important spectral confusion should attract more attention: there were few areas of bare soil, and thus only 14 pixels were selected as test samples during the visual interpretation. However, all these bare soil pixels were misclassified as vegetation and dark impervious surfaces, which indicates that the spectral signature of bare soil is very closed and can be confused with the spectral signature of vegetation and dark impervious surfaces. The overall accuracy of the ANN result was 85.52% and the Kappa coefficient was 0.8091. In the SVM classification, dark impervious surfaces were also misclassified as vegetation (51 pixels) and bare soil (35 pixels). Bright impervious surfaces were misclassified as vegetation (18 pixels) and dark impervious surfaces (13 pixels). In addition, the confusion between bare soil and vegetation and dark impervious surface also had a significant negative impact. The overall accuracy was 87.21% and the Kappa coefficient was 0.8288.

5.3.4 Cape Town

The LULC classification in Cape Town shows more complex spectral confusion between various land cover types. Figure 5.4 illustrates the classification results using ANN and SVM. With a visual interpretation of the results, an overall overestimation of bare soil can be observed from the SVM result. However, unlike the results in Guangzhou, Mumbai, and Sao Paulo, the spectral confusion was much more complicated since it occurred between different pairs of land covers. For instance, a number of vegetation pixels and a number of impervious surface pixels were both wrongly classified as bare soil pixels in the SVM result. The best identification of land cover type were water surfaces, with a rather consistent classification result from both ANN and SVM.

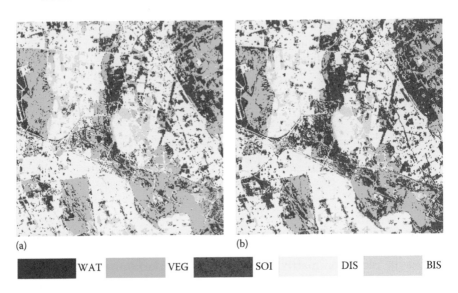

(a)   (b)

WAT   VEG   SOI   DIS   BIS

FIGURE 5.4
LULC classification in Cape Town: (a) ANN and (b) SVM.

In order to better investigate the spectral confusion between various land cover types, the confusion matrix was calculated using the testing samples. As shown in Tables 5.7 and 5.8, using the ANN approach, 24 pixels of water surfaces were misclassified as vegetation, while 11 pixels of vegetation were incorrectly classified as bare soil. Bare soil was misclassified as vegetation (33 pixels) and dark impervious surfaces (21 pixels). Dark impervious surfaces were misclassified as bare soil (27 pixels) and bright impervious surfaces (26 pixels). In addition, 11 pixels of bright impervious surfaces were classified as bare soil. This complicated spectral confusion was produced by the complex land covers in this study area. In Cape Town, there was dark vegetation from forest and bright vegetation from farmland and grassland. Dark vegetation was easily confused with water surfaces in some wetland. Bare soil in Cape Town is also diverse, including land under construction and seasonal bare

TABLE 5.7

Confusion Matrix of ISE in Cape Town
Using ANN

	WAT	VEG	SOI	DIS	BIS
WAT	74	0	0	3	0
VEG	24	201	33	5	1
SOI	1	11	203	27	11
DIS	3	0	21	221	8
BIS	0	0	13	26	179

Note: Overall accuracy: 82.44%, Kappa coefficient: 0.7756.

TABLE 5.8

Confusion Matrix of ISE in Cape Town
Using SVM

	WAT	VEG	SOI	DIS	BIS
WAT	69	0	0	2	0
VEG	26	163	12	5	1
SOI	5	49	234	28	16
DIS	2	0	23	228	37
BIS	0	0	1	19	145

Note: Overall accuracy: 78.78%, Kappa coefficient: 0.7261.

soil that should be covered by grass in the wet season. This bare soil can be easily confused with dark impervious surfaces, which are mainly from heavily residential areas. When using SVM, these situations can also be observed. There were 26 water pixels classified as vegetation and 49 vegetation pixels classified as bare soil. Bare soil was misclassified as vegetation (12 pixels) and dark impervious surfaces (23 pixels), while bright impervious surfaces were misclassified as bare soil (16 pixels) and dark impervious surfaces (37 pixels).

5.4 Conclusion

This chapter provided an assessment of the urban land cover complexity by analyzing the spectral confusion between various land cover classes in the LULC classification results using two popular methods, ANN and SVM. The Landsat images from Guangzhou, Mumbai, Sao Paulo, and Cape Town were selected according to the availability of corresponding SAR data. Five land cover types—water, vegetation, bare soil, dark impervious surfaces, and bright impervious surfaces—were classified from each of the Landsat images. Experimental results indicate some useful findings about the urban land cover complexity in tropical and subtropical areas. Firstly, ANN and SVM showed different behaviors in different study areas. In Guangzhou and Sao Paulo, the classification results are better from SVM than ANN, showing it is less influenced by the spectral confusion between bare soil and other land cover types. Meanwhile, in Mumbai and Cape Town, ANN was superior to SVM in eliminating the confusion between bare soil and other land cover types. Additionally, land cover confusion is very complex in tropical and subtropical urban areas compared with previous studies reported in the literature. Bare soil is the most likely land cover type to be confused with other land covers, including not only bright impervious surfaces, but also dark impervious surfaces and vegetation. Moreover, in Guangzhou, Mumbai, and Sao Paulo, the spectral confusion happens mainly in between one or two pairs of land cover classes, which is simpler than the case in Cape Town where the spectral confusion occurs between several pairs of land covers at the same time. Therefore, the spectral confusion in urban land covers is much more complicated in tropical and subtropical areas compared with the situations in temperate regions, which can have a fundamental impact on ISE in these regions. In the following chapters, SAR images will be synergized together with optical images to reduce the spectral confusion in the optical images.

6

Comparative Studies with Different Image Data and Fusion Methods

6.1 Comparison of ISE with Single Optical and SAR Data

This section aims to compare single optical data and single SAR data in terms of ISE to investigate the potential of SAR data in mapping impervious surfaces. Four sets of optical and SAR images in four different cities were employed in this section: Guangzhou, Mumbai, Sao Paulo, and Cape Town, which are located in tropical and subtropical regions. Landsat TM/ETM+ images and ENVISAT ASAR images in these cities were used as the optical and SAR images for the comparative experiment. Table 6.1 lists the optical and SAR images used in this section.

6.1.1 Parameter Configurations of ANN and SVM for Optical Data

ANN and SVM were employed as the methods to extract impervious surfaces from both optical and SAR images. The working principles of ANN and SVM were given in Chapter 2. Successful application of ANN and SVM lies in the parameter configurations regarding the datasets. As discussed in Chapter 2, the BP algorithm is crucial to the ANN classifier, while the number of iterations is a key factor for the classification. Therefore, in this study, the iteration times were changed from 1 to 1200. For each time period, we kept other parameters unchanged. For the SVM, two very important parameters were identified previously: the penalty (C) parameter and the Gamma (G) parameter in the kernel function.

Taking Guangzhou as an example, the impacts of various parameters on the accuracy of ISE are shown in Figure 6.1. Due to the characteristics of the Landsat data for the Guangzhou site, we tested different values of Gamma, including 0.1, 1.0, and 10, and found that the classification accuracy changed only a little. As a result, the penalty (C) must be quantitatively analyzed. For the ANN classifier, the following characteristics can be observed (Figure 6.1): (1) the accuracy fluctuated from 0 to 400 times and reached the peak (overall accuracy: 91.03%, Kappa coefficient: 0.88) at 400 iteration times, (2) from

TABLE 6.1

Datasets of Four Cities for Comparing Optical and SAR Images

Study Site	Optical Image	SAR Image
Guangzhou	Landsat ETM+	ENVISAT ASAR
Mumbai	Landsat TM	ENVISAT ASAR
Sao Paulo	Landsat TM	ENVISAT ASAR
Cape Town	Landsat TM	ENVISAT ASAR

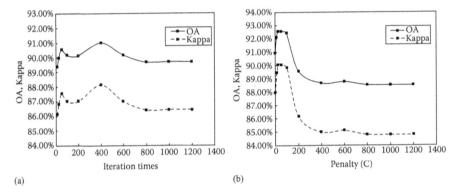

FIGURE 6.1

Impacts of parameter configuration for Landsat imagery: (a) ANN classifier and (b) SVM classifier (Gamma = 0.1).

400 iterations, both the overall accuracy and the Kappa coefficient decreased as the iteration times increased, indicating an overfitting of the BP learning algorithm, and (3) the accuracy finally became stable after 800 times, and the overall accuracy was about 89.7%, with a Kappa coefficient of 0.86.

On the other hand, results from SVM show that (1) at the first stage, the accuracy increased as the penalty increased and reached the highest point (overall accuracy: 92.55%, Kappa coefficient: 0.9) when the penalty was 50, (2) when the penalty was between 50 and 100, the accuracy stayed at a relatively stable level, and then decreased sharply to a lower level (overall accuracy: 88.66%, Kappa coefficient: 0.85), and (3) after the penalty reached 400, the accuracy reached another stable level as the penalty increased. In this way, the optimal parameters could be found for other study sites.

6.1.2 Parameter Configurations of ANN and SVM for SAR Data

As in the above section, by using the Guangzhou images as an example, the parameter configurations were also applied to SAR data to select the optimal parameters. Compared with the case of Landsat data, patterns were totally different for ASAR images (Figure 6.2). First, for ANN, the accuracy fluctuates frequently as the number of iterations increases. Before 200 iterations, the

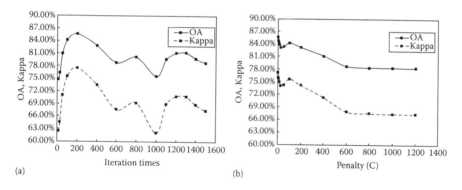

(a)   (b)

FIGURE 6.2
Effects of parameter configuration for ASAR imagery: (a) ANN classifier and (b) SVM classifier (Gamma = 0.1).

overall accuracy jumped from 74.75% to 85.66%, with an increase of Kappa coefficients from 0.62 to 0.78. The accuracy then underwent a stable decrease between 200 and 600 iterations, where the overall accuracy was 78.76% and the Kappa coefficient was 0.68. Next, the accuracy began to fluctuate as the iteration time increased. However, the highest accuracy within this period did not exceed that at 200 iterations. This trend can be illustrated by the ANN training RMS seen in Figure 6.3, where the RMS shows a relatively stable fluctuation when the iteration times were between 1000 and 4000.

Second, although there was a small fluctuation between the penalty of 1 and 100, an overall decrease can be observed toward both the total precision and Kappa coefficient when the penalty parameter increased from 1 to 1200. Therefore, the best accuracy came when the penalty was 1, where the overall accuracy was 85.53% and the Kappa coefficient was 0.77. As the penalty increased from 100 to 600, the accuracy underwent a linearly stable decrease

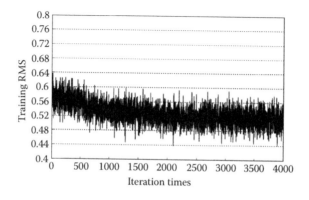

FIGURE 6.3
ANN training RMS with 4000 iterations.

to the point of a total precision of 78.68 and a Kappa coefficient of 0.68. The accuracy then remained nearly unchanged as a higher penalty was assigned.

6.1.3 Comparative Results of ISE

6.1.3.1 Guangzhou

Using the optical images, impervious surfaces were classified with MLP and SVM. The classification results with MLP demonstrate some important characteristics. First, an overall overestimation of bare soil in the study area can be observed, where some of the bright impervious surfaces in the central part of the area were incorrectly classified as bare soils. Second, shade, produced by clouds, hills, trees, and buildings, were mainly treated as water bodies and confused with dark impervious surfaces, especially in the northwest part in the study case. Third, cloud coverage was another factor decreasing the accuracy of classification. In contrast, results with SVM had some significant differences. Figure 6.4a and b shows that bare soil in the study area was classified more correctly and thus less bright impervious surfaces were identified. Moreover, the influence of shade was also not very great since more impervious surfaces in the shade area were successfully identified. The same difficulties were encountered with SVM with regard to clouds.

Compared with the classification of ETM+ images, the classified results from the ASAR images (Figure 6.4c and d) show some very different characteristics. First, only three classes could be obtained from the ASAR images: impervious surfaces, vegetation, and water bodies. Second, due to the interaction between microwaves and surface objects, some reflectance from the objects along the river banks and boats in the rivers appear in the ASAR images and thus produced some errors to the classifications. Third, ANN appears to be more suitable for the classification of ASAR images because there is much less noise and the edge effect is also much less. However, SVM encounters numerous noises (small areas that are wrongly recognized as water), and the edge effect is also obvious.

From the analysis of the classification accuracy about ETM+ and ASAR imagery using ANN and SVM, we can get a general idea of the comparison between these two different data sources for ISE. On the one hand, Landsat ETM+ images provide more information for ISE since there are multiple spectral bands. However, as an optical sensor, Landsat ETM+ tends to be impacted by clouds, and this situation becomes much more serious when the study areas are located in a humid area such as the PRD. On the other hand, ENVISAT ASAR images are able to provide more information in humid and work in all weather conditions. Nevertheless, ENVISAT ASAT cannot obtain many spectral bands, which leads to the overall low accuracy of ISE.

Additionally, methodology is another significant factor influencing ISE. In this study, two classifiers, ANN and SVM, are selected to compare their performances of ISE. With the above experiment and analysis, the best parameter

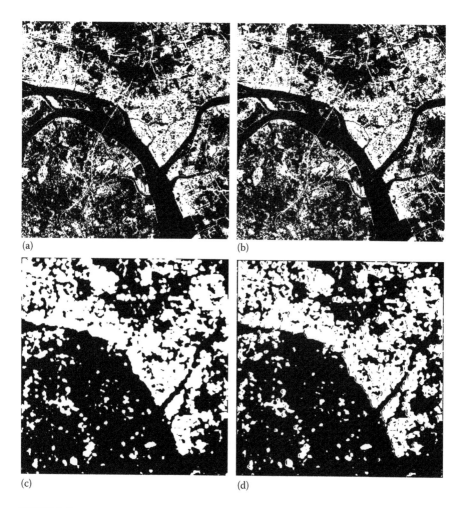

FIGURE 6.4
ISE using optical and SAR images alone in Guangzhou. (a) Landsat TM using ANN (December 31, 2010), (b) Landsat TM using SVM (December 31, 2010), (c) ENVISAT ASAR using ANN (December 23, 2010), and (d) ENVISAT ASAR using SVM (December 23, 2010).

configurations are chosen for the comparison of these two methods. Table 6.2 compares the differences of classification accuracy of ANN and SVM classifiers. It is impressive that using single ETM+ data alone gets a higher accuracy than using ASAR data alone. Moreover, SVM tends to be superior over ANN for the classification of Landsat ETM+ images, which is consistent with results reported in Sun et al. (2011). Nevertheless, SVM appears to be inferior to ANN when applied to ASAR images, which is quite an interesting result in the estimation of impervious surfaces. The results are also significant and meaningful for further research into the synergistic use of optical and SAR data to more accurately estimate impervious surface distribution.

TABLE 6.2

Impervious Surface Estimation in Guangzhou

Datasets	ANN		SVM	
	Overall Accuracy	Kappa Coefficient	Overall Accuracy	Kappa Coefficient
Landsat ETM+	91.03%	0.8813	**92.55%**	**0.9006**
ENVISAT ASAR	**85.66%**	**0.7754**	85.53%	0.7718

6.1.3.2 Mumbai

Figure 6.5 demonstrates the impervious surfaces estimated from only an optical image and only a SAR image using ANN and SVM classifiers. Some features can be identified from the ISE results by comparison. First, the results from only optical images shown in Figure 6.5a and b illustrate that there are both advantages and disadvantages in using single optical images. On the one hand, land cover boundaries between urban areas and vegetation or water surfaces were well identified with the rich multispectral images. On the other hand, a general overestimation can be observed over the whole study area due to the fragmentation of the urbanization process. For instance, some vegetated areas were misclassified as dark impervious surfaces, while some bare soil land was incorrectly identified as bright impervious surfaces. Additionally, by comparing the results from ANN and SVM, no significant difference can be observed from Figure 6.5a and b, indicating consistent results from ANN and SVM. Second, Figure 6.5c and d present the ISE results derived from single SAR images, showing a significantly different pattern from that of the optical images. The impervious surface distribution demonstrates an overall result that is not as good as the one from optical images. Due to the speckles in SAR images, boundaries between different land cover types are confused and mixed together. Impervious surfaces in central urban areas were generally underestimated. Some misclassification can be found in the water surfaces, which was caused by the high backscattering signals from the sea waves. However, there are some advantages that we can see from the results. For instance, vegetation and bare soil land were well classified and separated from other land covers because of their low backscatter coefficients in the SAR images. In comparing ANN and SVM, the results are consistent as in the optical image case.

Therefore, there are both advantages and disadvantages to optical and SAR images for ISE, even though the results from optical images are generally better. The critical observation from these comparative results is that the advantages and disadvantages are different in the results from optical images and from SAR images. This indicates that both of the datasets may be able to provide complementary information to each other for a better ISE

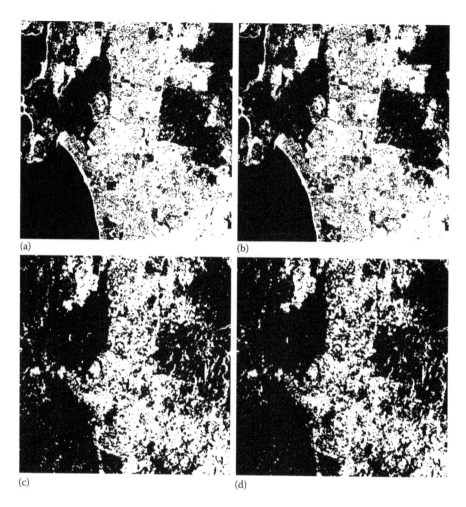

FIGURE 6.5
ISE using optical and SAR images alone in Mumbai. (a) Landsat TM using ANN (May 3, 2010), (b) Landsat TM using SVM (May 3, 2010), (c) ENVISAT ASAR using ANN (June 27, 2010), and (d) ENVISAT ASAR using SVM (June 27, 2010).

compared to using single dataset. For instance, on the one hand, vegetation and bare soils cannot be well classified using only optical images, but they can be identified very well using optical images due to their low backscattering characteristics. On the other hand, water surfaces cannot be classified correctly in SAR images because of the speckles from the water surface waves, while they are well classified in optical images because of their low reflectance in the multispectral bands.

Table 6.3 shows the accuracy assessment of the ISE results from optical and SAR images in Mumbai, which are consistent with the classified images in Figure 6.5. First, accuracy of the ISE result from the optical images was

TABLE 6.3

Accuracy Assessment of Impervious Surface Estimation in Mumbai

	ANN		SVM	
Datasets	Overall Accuracy	Kappa Coefficient	Overall Accuracy	Kappa Coefficient
Landsat TM/ETM+	88.59%	0.7708	87.14%	0.7421
ENVISAT ASAR	75.57%	0.4997	73.62%	0.4559

generally higher than that from the SAR images. The overall accuracy from the Landsat TM images was over 87% and the Kappa coefficient was over 0.74. However, using the SAR images alone, the overall accuracy was about 73%~75% and the Kappa coefficient was about 0.46~0.50, showing that SAR data alone cannot provide as a good result as optical data alone. Second, the accuracy comparison of ANN and SVM shows that ANN obtained slightly better results than SVM in both cases using optical data and SAR data. One possible reason for this may be the LULC complexity, which was discussed in Section 5.3.2. The spectral confusion mainly occurred between vegetation and impervious surfaces, as well as between bare soil and impervious surfaces. In this case, ANN showed a better result with the statistical risk minimization approach than SVM. Nevertheless, the difference of accuracy between ANN and SVM was not large and thus classification methods are not the major issue in this study case.

6.1.3.3 Sao Paulo

The impervious surfaces classified from single optical and SAR images are shown in Figure 6.6. Similarly, ANN and SVM were comparatively employed in the Sao Paulo case. Compared with the results in Guangzhou and Mumbai, the Sao Paulo case provides a relatively different view of the ISE results not only in using different datasets but also in using different classification methods. First, using single optical images, impervious surfaces were generally well classified over the whole study area without significant overestimation. However, some misclassification can still be found in Figure 6.6a and b. For instance, nearly the whole airport in the eastern part of the area was classified as impervious surfaces while in reality there were mostly vegetated areas and bare soil areas at the airport. Spectral confusion between bare soil and impervious surfaces also affected the results in the western part of the study area. Moreover, ANN and SVM showed a different estimation of impervious surfaces in this case. SVM got a better result with less impact from the spectral confusion between vegetation and impervious surfaces and between bare soil and impervious surfaces, especially in the western part of the study area. Second, the ISE result from a single SAR image also provided some unique characteristics. Affected by the speckles

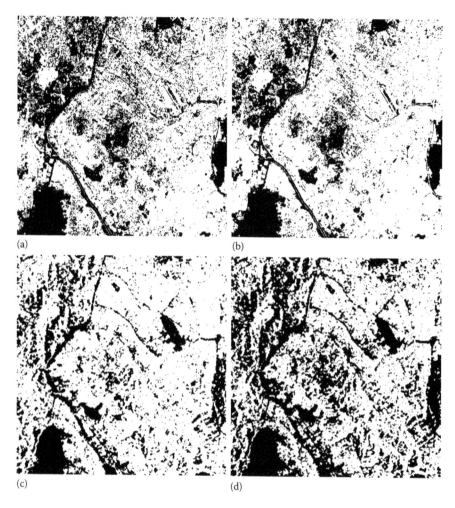

FIGURE 6.6
ISE using optical and SAR images alone in Sao Paulo. (a) Landsat TM using ANN (February 5, 2010), (b) Landsat TM using SVM (February 5, 2010), (c) ENVISAT ASAR using ANN (January 5, 2010), and (d) ENVISAT ASAR using SVM (January 5, 2010).

in the SAR image, linear features such as roads and rivers could not be well classified, leading to the misclassification on the boundaries between different land covers. Moreover, SAR data in the Sao Paulo case did not show good ability in identifying vegetated areas. Even though most of the airport area could be well classified as nonimpervious surface, other vegetation in the residential regions located in the middle and western part could not be well classified. This result was actually caused by the tall buildings in these areas. Tall buildings, although located among high vegetated greening areas, can cause high a backscattering coefficient in the SAR images. Consequently, SAR images mistakenly recognized them as impervious surfaces even though they are located among vegetated areas. Moreover, ANN and SVM again showed different behavior using only SAR images. ANN seemed to overestimate the impervious surfaces by incorrect classification of vegetation over the whole area. However, SVM was influenced by the speckles in SAR images and thus produced noisy impervious surfaces mapping in the result. Underestimation of impervious surfaces can also be observed in the central urban areas by the SVM classifier. Comparatively, the ISE experiment in Sao Paulo also indicates that there are both advantages and disadvantages of optical and SAR images for ISE. This also indicates that both of the datasets may be able to provide complementary information to each other for a better ISE compared to using single dataset alone. Therefore, it is necessary and practical to fuse both the optical and SAR images at the same time to improve the accuracy of ISE.

Accuracy assessment of the ISE results from optical and SAR data in Sao Paulo is shown in Table 6.4, providing a similar result to those in Figure 6.6. The overall accuracy from the Landsat TM images was over 87% and the Kappa coefficient was over 0.74. However, using the SAR image alone, the overall accuracy was only about 73%, which was 14% lower than that using the optical image alone. The Kappa coefficient was about 0.46. This indicates that SAR data alone cannot provide the same good results as can optical data alone. In addition, an accuracy comparison of ANN and SVM shows that SVM obtained better results than ANN in both cases using optical data and SAR data. This result is different from that in Mumbai. When considering the LULC complexity in Sao Paulo outlined in Section 5.3.3, we find that spectral confusion also happened between vegetation and impervious surfaces, and between bare soil and impervious surfaces. As indicated by the result in Mumbai, ANN should perform better than SVM in Sao Paulo. However, when we compare the number of training samples in the two cases, we find that Sao Paulo has fewer training samples (545 pixels) than Mumbai (483 pixels). Although ANN can obtain better results with a statistical principle in a complex LULC environment, this advantage depends on a sufficient number of samples. When there are not enough samples, SVM can do better with a structural risk minimization approach. Nevertheless, considering the accuracy, the difference between ANN and SVM is not significant and thus classification methods are still not a major issue in this case.

TABLE 6.4

Accuracy Assessment of Impervious Surface Estimation in Sao Paulo

	ANN		SVM	
Datasets	Overall Accuracy	Kappa Coefficient	Overall Accuracy	Kappa Coefficient
Landsat TM/ETM+	87.21%	0.7453	88.99%	0.7805
ENVISAT ASAR	72.82%	0.4579	73.18%	0.4671

6.1.3.4 Cape Town

The estimation of impervious surfaces using an optical image alone and a SAR image alone is demonstrated in Figure 6.7. ANN and SVM were comparatively applied for ISE in the Cape Town study area. Since the landscape of Cape Town is unique with relatively separated impervious surfaces and nonimpervious surfaces, it is easier to understand the results. From the ISE result derived from an optical image alone, a small amount of impervious surfaces were incorrectly classified as bare soil land and vegetated areas, which was mainly caused by the spectral confusion between bare soil and

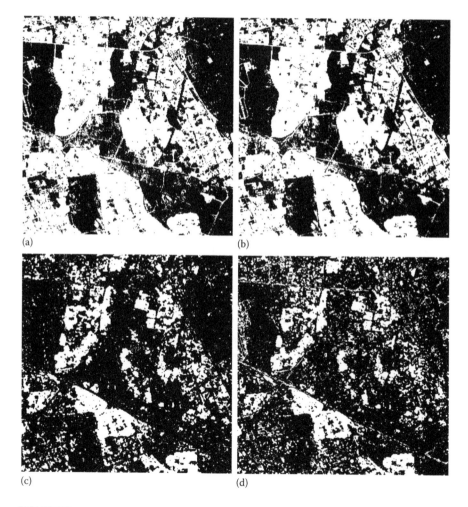

(a)   (b)   (c)   (d)

FIGURE 6.7
ISE using optical and SAR images alone in Cape Town. (a) Landsat TM using ANN (June 2, 2011), (b) Landsat TM using SVM (June 2, 2011), (c) ENVISAT ASAR using ANN (August 9, 2011), and (d) ENVISAT ASAR using SVM (August 9, 2011).

impervious surfaces and between vegetation and impervious surfaces. However, linear features such as the road network can be identified clearly in the Cape Town region. Nevertheless, the ISE results from the SAR images alone in Cape Town were rather worse compared with those from only optical images. First, too many noises are noticeable over the whole area caused by the speckles in the SAR image. Second, impervious surfaces were greatly underestimated, with large area of impervious surfaces misclassified as nonimpervious surfaces. Third, most of the roads failed to be identified over the study area. These three negative effects were caused by the speckles in the SAR images. Therefore, the comparison indicates SAR data alone cannot provide good ISE results compared to using optical data alone. Moreover, since there are two speckle noises in the result images, it is difficult to identify the effectiveness of SAR data in separating impervious surfaces and vegetation or bare soil. Therefore, more assessment should be done through quantitative analysis of the ISE results.

The accuracy assessment of the ISE results (Table 6.5) shows that optical data alone provided much higher accuracy than that from SAR data alone. Using only a Landsat TM/ETM+ image, the overall accuracy was 92.11% and the Kappa value was 0.8405 with the ANN classifier. The overall accuracy was 92.68% and the Kappa value was 0.8514 with the SVM classifier. In contrast, using only the ENVISAT ASAR image, the overall accuracy is 64.32% for the ANN classifier and 63.00% for the SVM classifier, with a Kappa coefficient of 0.2386 and 0.2196, respectively. Therefore, the accuracy from single SAR data is rather low compared with the accuracy from optical data. This is consistent with the results demonstrated in the ISE images in Figure 6.7. In terms of the classification methods, ANN and SVM did not show significant differences in the results using either dataset, and thus it is not the major issue compared with the dataset itself.

6.1.4 Discussion and Implications

This study compares optical and SAR data in terms of estimation of impervious surfaces using a single data source. Experimental results in four different cities of the tropical and subtropical regions show some important findings

TABLE 6.5

Accuracy Assessment of Impervious Surface Estimation in Cape Town

Datasets	ANN		SVM	
	Overall Accuracy	Kappa Coefficient	Overall Accuracy	Kappa Coefficient
Landsat TM/ETM+	92.11%	0.8405	92.68%	0.8514
ENVISAT ASAR	64.32%	0.2386	63.00%	0.2196

for both the advantages and disadvantages of each data source. First, in all the cases, using optical images alone provided a generally better result than using SAR data alone. The difference of overall accuracy varied from about 7% to about 29%, while the difference of Kappa coefficient varied from about 11% to about 60%. The results demonstrate that using optical images alone provide generally better identification of vegetation, dark impervious surfaces, and bright impervious surfaces, even though there are spectral confusions between impervious surfaces and vegetation or bare soils. However, due to speckles of SAR images, the ISE results using SAR data alone were affected by numerous noises, especially on the boundaries between different land covers. These noises can influence the classification results dramatically and lower the accuracy, depending on the complexity of land cover patterns. In particular, linear features such as roads and bridges cannot be correctly identified using SAR data only. Second, SAR data was able to show some advantages for ISE compared with optical data. For instance, separation between bright impervious surfaces and bare soils could be reduced due to their different backscattering behaviors with microwave remote sensing. In addition, spectral confusions between dark impervious surfaces and vegetation could be reduced to some extent in the SAR images. Therefore, optical images and SAR images can provide complementary information for each other to improve the estimation of impervious surfaces. Third, in comparing the ANN and SVM classifiers, both methods demonstrated similar results when applied to the same dataset in the same study area. The differences of accuracy between the results from ANN and SVM were less than 1%. In general, our experimental results showed that SVM was more appropriate for using optical data alone, while ANN provided better results when using SAR data alone. However, this parameter is not so strong for all cases and it should depend on the land cover diversity in a specific application.

6.2 Comparison of Different Levels of Fusion Strategies

For combining optical and SAR data to extract impervious surfaces, selection of an appropriate level of fusion (e.g., pixel level, feature level, or combinational level) remains unclear. This section aims to address the issue by comparing different schemas of fusion strategies and exploring the best choices between optical and SAR remote sensing to improve the accuracy of ISE. The implementation of fusion strategies at different levels is just a general one that aims at investigating the overall differences of various strategies. In this section, SVM, one of the most commonly used machine-learning models, is employed to conduct the fusion at different levels. Additionally, since the comparison of fusion levels can be influenced by many factors such as

TABLE 6.6

Datasets of Four Cities for Comparing Different Fusion Levels

Study Site	Optical Image	SAR Image
Shenzhen	SPOT-5	ENVISAT ASAR
Mumbai	Landsat TM	TerraSAR-X
Sao Paulo	Landsat TM	TerraSAR-X
Cape Town	Landsat TM	ENVISAT ASAR

working frequency of SAR, spatial resolutions of optical and SAR data, and land cover diversity, this section does not use the same datasets as in Section 6.1; instead, various combinations of optical and SAR data with different sensors and spatial resolutions are selected to compare the impacts of different fusion levels. Since SPOT-5 data was not available, another city, Shenzhen, was chosen instead of Guangzhou by employing SPOT-5 and ENVISAT ASAR data. Table 6.6 shows the optical and SAR images in four different cities for this comparative experiment. There are different combinations of optical and SAR images; for instance, SPOT-5 and ENVISAT ASAR data were used for the Shenzhen case, Landsat TM and TerraSAR-X data were used for study the Mumbai and Sao Paulo cases, and Landsat TM and ENVISAT ASAR were used for the study in the Cape Town case. These various combinations of datasets could also test the impacts of different spatial resolutions of optical and SAR images and different working modes of SAR images.

6.2.1 Feature Extraction from Optical and SAR Data

The extracted features were prepared for the feature level fusion and decision level fusion. Two groups of features were extracted from the optical and SAR images, respectively. The first feature group, from optical images, included spectral features and texture features. In this study, two popular indices, NDVI and NDWI, were applied and considered as the spectral features, since they are calculated from the multispectral bands of optical images. For the texture features, GLCM was applied to the optical images with the size of the moving window as 3×3 and 7×7 according to our previous study. The 3×3 pixel moving window was used to calculate the GLCM texture features for Landsat TM images and the corresponding resampled SAR images with lower spatial resolution. The 7×7 pixels window was applied to the SPOT-5 and corresponding resampled SAR images with higher spatial resolution.

6.2.2 Fusion Strategies at Different Levels

A number of methods were proposed to synergistically use optical and SAR images for various applications. Contextual information from neighboring pixels was considered to be useful and Markov random fields (MRFs) was used for the combined use of optical and SAR data (Solberg and Jain 1997).

TABLE 6.7

Fusion Strategies of Three Different Levels

Fusion Level	Optical Image	SAR Image
Pixel level	Original multispectral bands data	Backscattering intensity
Feature level	NDVI, NDWI, GLCM features	GLCM textures
Combinational level	Multispectral bands, GLCM features	Backscattering intensity, GLCM textures

Nonparametric approaches (e.g., ANN and SVM) were also applied with the two data sources concatenated in a stacked vector (Pacifici et al. 2008). However, optical and SAR data should be treated differently since they carry different kinds of information (Tupin and Roux 2005). Thus, more advanced methods should be used by treating optical and SAR data in a different way. Decision fusion methods are used as they make final decisions on the results from different sources of data (e.g., by weighting the influence of different multisensors) (Pacifici et al. 2008). Additionally, ensemble methods (e.g., RF) are also applied at the decision level (Waske and van der Linden 2008). For the estimation of impervious surfaces, some research is conducted based on the use of optical and SAR images (Jiang et al. 2009; Leinenkugel et al. 2011; Yang et al. 2009b).

However, only parts of information (e.g., coherence, average intensity, temporal change of intensity) were considered, and further research is needed to comprehensively explore the potentials of SAR data and the synergistic use of optical and SAR data for the accurate estimation of impervious surfaces. Moreover, the differences of different levels of fusion (e.g., pixel level, feature level, and decision level) remain unknown in terms of ISE.

In order to compare different levels of fusion strategies, the pixel level, feature level, and decision level are designed in this section. For the three levels of fusion, SVM is applied as the fusion method. To apply SVM, data from both the SPOT-5 and ASAR images are put together as the input to SVM, and the output is binary with impervious and nonimpervious surfaces. At the pixel level, the SPOT-5 and ASAR images are treated as the input without any feature extraction operations. At the feature level, texture features are extracted from both the SPOT and ASAR images, and these texture feature maps will be the input of the fusion instead of the data at the pixel level. For the combinational level, both the original multispectral bands and GLCM-based texture features were combined for the ISE. Table 6.7 shows the data or features from the two data sources in different levels of fusion.

6.2.3 Fusion Results on Different Levels

6.2.3.1 Shenzhen

Fusion between SPOT-5 and ASAR images was conducted in the Shenzhen case at three levels: the pixel, feature, and combinational levels. Figure 6.8 reveals the results of ISE from the three different levels of fusion. A false

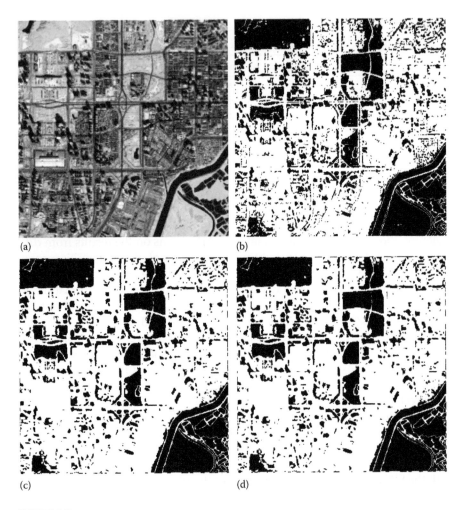

FIGURE 6.8
ISE with optical SAR fusion in Shenzhen. (a) SPOT-5 (RGB: 4-1-2), (b) pixel-level fusion, (c) feature-level fusion, and (d) combinational-level fusion.

color image of the study area (Figure 6.8a) is also provided to give a better understanding of the results. There are several interesting findings. First, the results from pixel level and feature level fusion were characterized with more shaded areas (dark holes) (Figure 6.8b and c) due to the shadows from tall buildings (Figure 6.8a). However, these dark holes are reduced in the result of combinational fusion (Figure 6.8d). Second, it shows some edge effects in the results of feature-level and decision-level fusions. These edge effects are caused by the use of texture features, as they are calculated with four window sizes. Third, bare soils can be more easily separated at the feature level and combinational level fusions. Located on the dark blocks of the result images are some bare soils, appearing in yellow and bright colors in the corresponding false-color image (Figure 6.8a). This shows us that these dark blocks are more pure in Figure 6.8c and d, while some small areas of bare soils were mistakenly classified as impervious surfaces in Figure 6.8b. To summarize, the result of decision fusion appears to be the best, as there are less shaded areas and less incorrectly classified bare soils.

6.2.3.2 Mumbai

The impervious surfaces estimated using combined optical and SAR images at various fusion levels are shown in Figure 6.9. For an easier understanding of the results, a false-color image of the Landsat TM data is provided in the

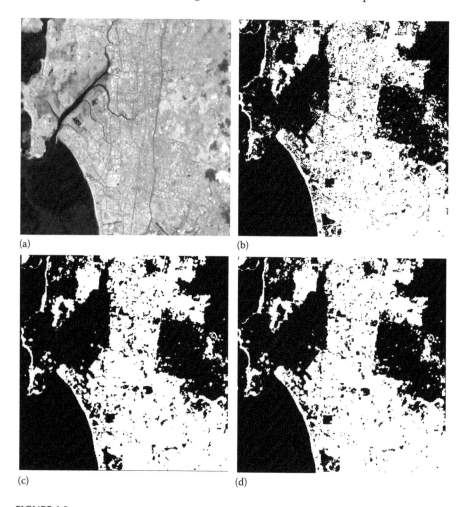

(a) (b)
(c) (d)

FIGURE 6.9
ISE with optical SAR fusion in Mumbai. (a) Optical image, (b) pixel-level fusion, (c) feature-level fusion, and (d) combinational-level fusion.

figure. At the pixel level, there was noticeable misclassification in vegetated areas and bare soils with some impervious surfaces wrongly identified. These incorrect impervious surfaces showed some characteristics such as noises that were actually caused by the speckles from SAR images. Due to the land cover characteristics, overestimation of impervious surfaces could be observed as in the result using a single optical image shown in Section 6.1. At the feature level, the influence from speckles in SAR images was reduced with less impervious surfaces classified from vegetation and bare soil areas. The boundaries between impervious surfaces and nonimpervious surfaces were also clearer, making the results generally better by visual interpretation. At the combinational level, compared with the results at the feature level, the noisy misclassified impervious surfaces from vegetation and bare soil were not reduced significantly. The boundaries between impervious surfaces and nonimpervious surfaces remained the same as at the feature level. However, the overestimation of impervious surfaces in the central urban areas was increased slightly. In general, feature-level and combinational-level fusion were more appropriate for the fusion between optical and SAR images for ISE in the Mumbai area. Pixel-level fusion was influenced by the speckles in SAR images and thus was not suitable in this study case.

6.2.3.3 Sao Paulo

The estimation of impervious surfaces using different levels of fusion with optical and SAR images is shown in Figure 6.10. The original optical image is provided in Figure 6.10a for a better visual interpretation of the ISE results. The results show some characteristics of ISE by combining optical and SAR images at various fusion levels. At the pixel level, influences of speckles in SAR data could be observed in intensive residential areas where tall buildings produced some shadows. As a result, some noisy impervious surfaces are noticeable on the upper part of the study area. Moreover, boundaries between impervious surfaces and nonimpervious surfaces were also unclear with noises. At the feature level, the estimated impervious surfaces were much better classified by visual interpretation. The speckle effect was noticeably reduced over the whole area. Boundaries between impervious surfaces and nonimpervious surfaces were much clearer. However, overestimation of impervious surfaces occurred slightly over some vegetated areas. At the combinational level, similar to the feature-level fusion result, the speckle effect was reduced and boundary classification was improved. Moreover, less overestimation can be found compared with the result at the feature-level fusion. Therefore, combinational-level fusion was the most appropriate strategy for combining optical and SAR images for ISE in the Sao Paulo case, while pixel-level fusion was not suitable due to the impact of the speckles from SAR images.

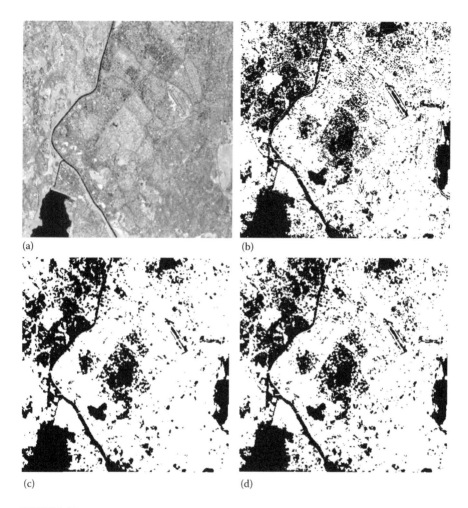

FIGURE 6.10
ISE with optical SAR fusion in Sao Paulo. (a) Optical image, (b) pixel-level fusion, (c) feature-level fusion, and (d) combinational-level fusion.

6.2.3.4 Cape Town

The results of impervious surfaces estimated from combining optical and SAR images at various fusion levels are shown in Figure 6.11. The first figure shows the original Landsat TM image for a better understanding of the results. It can be seen that impervious surfaces and nonimpervious surfaces are naturally separated from each other in a relatively regular way compared with other cities in this research. From the results, some differences can be noted for various fusion levels. At the pixel level, the speckles in the SAR image were able to influence the result with some noises in the vegetation

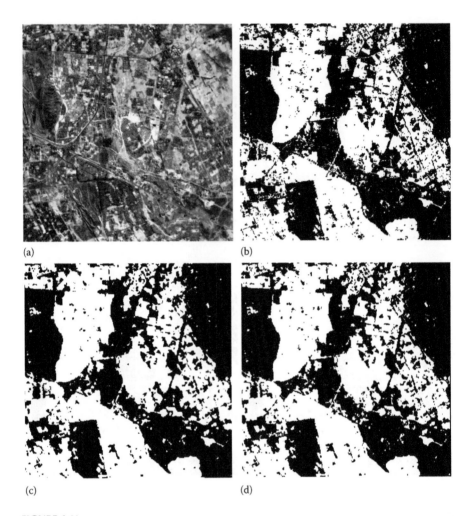

FIGURE 6.11
ISE with optical-SAR fusion in Cape Town. (a) optical image, (b) pixel level fusion, (c) feature-level fusion, and (d) combinational level fusion.

and bare soil land. Even though bare soil and vegetation are spectrally confused in the Cape Town area as shown in Chapter 5, this spectral confusion did not affect the result of ISE since vegetation and bare soil are both nonimpervious surfaces. This observation can be noticed in the results at all levels of fusion. At the feature level, the result looks better with clearer boundaries of impervious surfaces and nonimpervious surfaces, as well as cleaner vegetation and bare soil areas. The combinationa-level fusion result was very similar to that at the feature level. Both feature- and combinational-level fusion produced better ISE than pixel-level fusion. However, it should be noted that linear features such as road network were not well classified in all the levels of fusion, indicating the limitation of the additional use of SAR images in the Cape Town study area.

6.2.4 Comparisons of the Accuracy Assessment

In order to compare the results from different levels of fusion in a quantitative way, the overall accuracy and kappa coefficient were calculated based on the test samples. In addition, to evaluate the effectiveness of the fusion approach for ISE, an additional classification was applied on the optical images separately for each study case and corresponding accuracy was calculated for comparison. Tables 6.8 through 6.11 compare the accuracies over different strategies of ISE. Some interesting results can be observed. Different fusion levels illustrate different characteristics for combining optical and SAR data to estimate impervious surfaces. First, for all the four cases, pixel-level fusion did not show good ability to fuse the two data sources due to the decrease of the accuracy of ISE. Compared with using optical images alone, the decreasing accuracy at the pixel-level fusion are 0.20%, 0.32%, 4.09%, and 1.04%, respectively, for Shenzhen, Mumbai, Sao Paulo, and

TABLE 6.8

Comparison of Accuracy Assessment in Shenzhen

Fusion	Overall Accuracy	Kappa Coefficient
Single optical data	78.47%	0.5694
Pixel level	78.27%	0.5654
Feature level	**80.83%**	**0.6166**
Combinational level	80.53%	0.6107

TABLE 6.9

Comparison of Accuracy Assessment in Mumbai

Fusion	Overall Accuracy	Kappa Coefficient
Single optical data	92.23%	0.845
Pixel level	91.91%	0.8385
Feature level	**95.06%**	**0.901**
Combinational level	94.34%	0.8867

TABLE 6.10

Comparison of Accuracy Assessment in Sao Paulo

Fusion	Overall Accuracy	Kappa Coefficient
Single optical data	91.56%	0.8315
Pixel level	87.47%	0.7495
Feature level	92.40%	0.8481
Combinational level	**92.48%**	**0.8497**

TABLE 6.11

Comparison of Accuracy Assessment in Cape Town

Fusion	Overall Accuracy	Kappa Coefficient
Single optical data	92.68%	0.8514
Pixel level	91.64%	0.8305
Feature level	92.49%	0.8481
Combinational level	**93.05%**	**0.8596**

Cape Town, respectively. As a result, there is slight decrease in Shenzhen and Mumbai and a noticeable decrease in Sao Paulo and Cape Town. The inapplicability of fusing optical and SAR images at the pixel level has been reported previously (Soergel 2010; Zhang et al. 2010) indicating that the different working modes of optical and SAR remote sensing make pixel-level fusion inappropriate for combining the two sources of data. The results of this study support the conclusion found in the previous research. As well, we found that this decrease of accuracy is even more for urban areas in the southern sphere. Second, feature-level fusion could improve the accuracy of ISE compared with using optical data alone. The overall accuracy for the four cities were 78.47%, 92.23%, 91.56%, and 92.68% for Shenzhen, Mumbai, Sao Paulo, and Cape Town, and the Kappa coefficients were 0.5694, 0.845, 0.8315, and 0.8514, respectively. After fusing optical and SAR images, the overall accuracies were 80.83%, 95.06%, 92.40%, and 92.49%, and Kappa values were 0.6166, 0.901, 0.8481, and 0.8481. This shows a general improvement of the accuracy with the feature-level fusion of optical and SAR images. Third, combinational level fusion was also able to generally improve the ISE result; however, whether combinational-level fusion is superior to feature-level fusion depends on the study area. The overall accuracy with combinational level fusion are 80.53%, 94.34%, 92.48%, and 93.05% and the Kappa coefficients are 0.6107, 0.8867, 0.8497, and 0.8596. Hence, a general increase of both the overall accuracy and Kappa coefficient could be noticed. However, the improvement by feature-level fusion and combinational fusion is different for different cities and whether the combinational level fusion is better depends on the study area. For instance, the accuracy obtained by feature-level fusion was higher than that by combinational-level fusion in Shenzhen and Mumbai, while combinational-level fusion was superior in Sao Paulo and Cape Town.

These results may be caused by several factors, such as the resolution of the images, climate categories, and the land cover diversity of the study area. In Shenzhen, the optical data and SAR data are SPOT-5 and ENVISAT ASAR images, in Mumbai and Sao Paulo they are Landsat TM and TerraSAR-X images, and in Cape Town they are Landsat TM and ENVISAT ASAR images. Therefore, there is no direct relationship between the improvement of accuracy and resolution of optical and SAR images. It is interesting that TerraSAR-X is not necessarily better than ENVISAT ASAR data for improving ISE accuracy, although the resolution of TerraSAR-X data is much higher than that of

ENVISAT ASAR data. This may be caused by the downsampling of the SAR data before combining optical and SAR images, and thus the final results may depend on the resolution of the optical images. In terms of the climate types, Shenzhen and Sao Paulo have a subtropical humid climate, Mumbai has a tropical wet and dry climate, and Cape Town is located in a Mediterranean climate region. There is also no direct relationship between climate types and accuracy. However, when it comes to the land cover diversity of the study area, we find the land covers in Shenzhen and Mumbai are much more complicated than those in Sao Paulo and Cape Town. In Shenzhen and Mumbai, the urbanization process was more irregular and fragmentation of impervious surfaces could be observed in the study area. Nevertheless, the land covers in Sao Paulo and Cape Town are relatively simple with more clearly separated impervious surfaces and nonimpervious surfaces. To summarize, the results indicate that combinational-level fusion is more suitable for ISE of urban areas with less diverse land covers, while feature-level fusion is more appropriate for urban areas with more diverse land covers.

6.2.5 Discussion and Implications

This study compares three different fusion levels: pixel level, feature level, and combinational level, regarding optical and SAR images in terms of ISE. Spectral and texture features are extracted from the optical and SAR images, as well as the detailed design of fusion strategies for pixel-, feature-, and combinational-level fusions. SVM is then employed to conduct the fusion operation. The experimental results show some important conclusions for selecting the fusion strategy of fusing optical and SAR data. First, pixel-level fusion is not appropriate for optical and SAR image fusion due to speckles in SAR data (Soergel 2010). At the pixel-level fusion, there is no handling of the speckles, and thus this random signal can affect the fusion procedure and influence the final ISE results. Consequently, it reduces the accuracy even compared with the single use of optical data. This result is consistent in the four study cases. Second, both feature-level and combinational-level fusion are able to improve the accuracy of ISE by fusing optical and SAR data. However, whether feature-level or combinational-level fusion is better for optical-SAR fusion may depend on specific cases in terms of the land cover diversity and complexity. For instance, as the experimental results indicated, combinational-level fusion had best accuracy in Sao Paulo and Cape Town, where the land covers are not so complex and impervious surfaces and nonimpervious surfaces are more easily separated. Meanwhile, feature-level fusion had the best results in Shenzhen and Mumbai where much more complex land covers can be observed due to irregular urbanization with fragmentation of impervious surfaces. As a result, we conclude that combinational-level fusion is more suitable for ISE of urban areas with less diverse land covers, while feature-level fusion is more appropriate for urban areas with more diverse land covers.

6.3 Comparison of Different Image Features

6.3.1 Experiment Design

This comparison study aims to investigate the effectiveness of different image features of both optical and SAR images. Two study sites, Hong Kong and Sao Paulo, were selected with two different sets of optical and SAR images. In Hong Kong, SPOT-5 was used as the optical image and TerraSAR-X as the SAR image. In Sao Paulo, Landsat TM and TerraSAR-X were employed as the optical and SAR images, respectively. Table 6.12 shows the datasets used in this study.

In order to synergize the optical and SAR images, both spectral and spatial features were extracted from the two data sources. Table 6.13 shows the features extracted from the two images. For the optical image, NDVI and NDWI are calculated as the spectral feature and GLCM-based and SAN-based texture features are also calculated. Moreover, the shape feature based on SAN is also computed from the optical images. For the SAR images, as there is only one single band, the GLCM-based texture features are calculated. As described in Chapter 3, four texture measures, the mean, variance, HOM, DISS, ENT, and ASM, were employed as the GLCM-based texture measures, and the window size to calculate the GLCM was 3×3 pixels (for 30m resolution) and 7×7 pixels (for 10m resolution). To calculate the SAN-based features, the size of view port (Section 3.5, Figure 3.11) was set to be 11×11 pixels, ω_1, ω_2, and ω_3 in Equation 3.2 were empirically set to be 0.8, 0.2, and 0, respectively, and the H in Equation 3.3 was set as [1, 2] empirically. The form factor (F) in Equation 3.4 was chosen as the shape feature. Therefore, there were two texture features and two shape features of the SAN.

In terms of the fusion and classification methods, SVM was employed and compared to classify the land covers and estimate the impervious surfaces. Key parameters in these methods were optimized with an iteration process (see Section 6.1), where different values are tested and the ones with optimal

TABLE 6.12

Datasets of Two Cities for Comparing Different Features

Study Site	Optical Image	SAR Image
Hong Kong	SPOT-5	TerraSAR-X
Sao Paulo	Landsat TM	TerraSAR-X

TABLE 6.13

Selection of Feature Measures of Optical and SAR Images

Features	Optical Image	SAR Image
Spectral feature	NDVI, NDWI	None
Texture feature	GLCM texture, SAN texture	GLCM texture
Shape feature	SAN shapes	None

TABLE 6.14

Design of Combinational-Level Fusion

Code	Optical Image	Optical Image Features	SAR Image	SAR Features
OPT	Original data	N/A	N/A	N/A
OPT_SG	Original data	Spectral, GLCM	N/A	N/A
OPT_SGS	Original data	Spectral, GLCM, SAN	N/A	N/A
OPT_SAR	Original data	Spectral, GLCM	Intensity	GLCM
OPT_SARS	Original data	Spectral, GLCM, SAN	Intensity	GLCM

accuracy are used as the optimal parameters. The penalty (C) and gamma (G) in the kernel function are selected as the key parameters. The penalty was tested from 100 to 1500 with a step of 100, and the gamma was tested from 0.0000001 (10e-7) to 0.5 by testing 14 values.

For the fusion strategy, in this study, the combination level of fusion was adopted, as it takes into account both the original data and various features. In a combinational fusion, different combinations of features can be designed as the input of the fusion and classification. Table 6.14 demonstrates four different combination strategies where various combinations of optical features and SAR features are designed. There are five modes of strategies used in this study—OPT, OPT_SG, OPT_SGS, OPT_SAR, and OPT_SAR_S— with corresponding image data and features listed in Table 6.14. Among these modes, OPT includes only the original data of optical images. OPT_SG and OPT_SGS use only optical image data and features, while OPT_SAR and OPT_SARS use both optical and SAR images and their features.

6.3.2 Results of Feature Extractions

6.3.2.1 Features of Optical Images

6.3.2.1.1 Spectral Features

Spectral features were calculated from both SPOT-5 and Landsat TM images to represent the characteristics of spectral reflectance in each pixel. NDVI and NDWI were employed as the spectral features. Figure 6.12 shows the NDVI and NDWI of optical images in Hong Kong and Sao Paulo. It can be seen that vegetation was well recognized by high values of NDVI, and the water surfaces are highlighted in the NDWI images. Figure 6.12 shows that vegetated areas are mainly distributed on the mountainous regions and the greening areas of the residential region. Water surfaces are mainly located on the sea, rivers, and lakes. However, shaded areas also have high values of NDWI due to their low reflectance.

6.3.2.1.2 GLCM Features

GLCM was calculated from optical images and then various texture measures were calculated from the GLCM to extract the texture features. In this study, in order to extract the texture features as much as possible, six

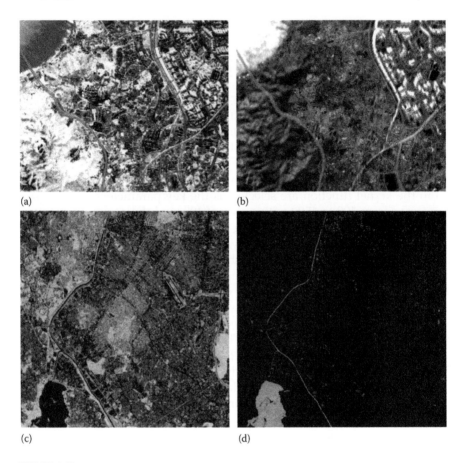

FIGURE 6.12
Spectral features of optical images in Hong Kong and Sao Paulo. (a) NDVI (Hong Kong), (b) NDWI (Hong Kong), (c) NDVI (Sao Paulo), and (d) NDWI (Sao Paulo).

frequently used texture measures were employed, including mean (MEA), variance (VAR), HOM, DIS, ENT, and SMA. Figures 6.13 and 6.14 show the GLCM-based texture feature images from the first band of the optical images in Hong Kong and Sao Paulo. These texture images illustrate different characteristics in high-resolution images (SPOT-5) and moderate-resolution images (Landsat TM).

First, in Hong Kong, water and land is well separated in all six feature images. Edges between two land covers are highlighted in the VAR feature. However, the sea area on the northwestern part is separated into two small parts caused by the water quality. Only the DIS image appears less impacted by the water quality difference. Moreover, some linear terrains (roads, highways, and rivers) also show some differences in each of the feature images. However, all these feature images illustrate an edge effect due to the rectangular windows when calculating GLCM.

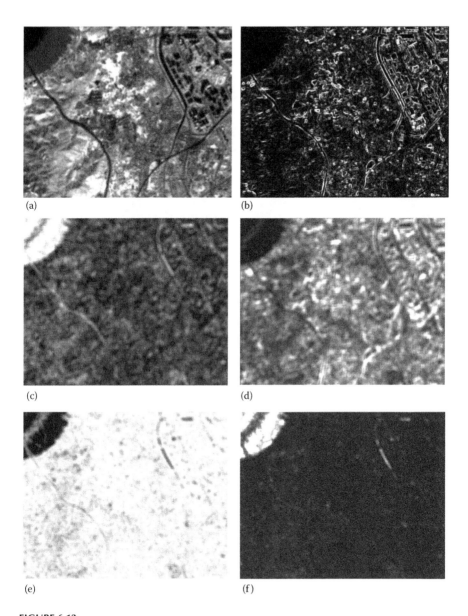

FIGURE 6.13
GLCM-based textures of SPOT image in Hong Kong. (a) MEA, (b) VAR, (c) HOM, (d) DIS, (e) ENT, and (f) SMA.

Second, the Sao Paulo images demonstrated a different view of these GLCM features. Water and other land covers could be separated well in the MEA and SMA features but rivers were only highlighted in MEA, HOM, and DIS features. The VAR image highlights the boundaries in the residential region and thus may be beneficial for classifying impervious surfaces from

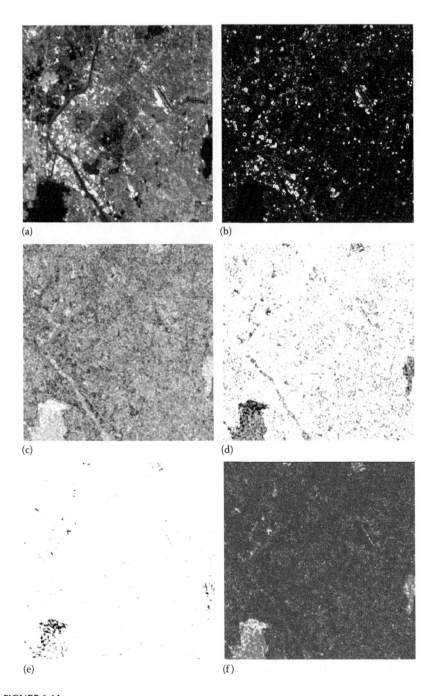

FIGURE 6.14
GLCM-based textures of SPOT images in Sao Paulo. (a) MEAM, (b) VAR, (c) HOM, (d) DIS, (e) ENT, and (f) SMA.

nonimpervious surfaces. Nevertheless, the ENT does not show too much useful information over the whole study area in Sao Paulo, which indicates that a feature selection procedure may be needed in order to select the most effective features before classification. The ineffectiveness of some GLCM texture features may be caused by two possible factors: the medium spatial resolution of the Landsat TM image and the land cover characteristics of Sao Paulo. Compared with the SPOT-5 image in Hong Kong, the lower resolution and simpler land cover characteristics in the Sao Paulo image show less rich texture information and thus lead to some ineffective texture feature images.

6.3.2.1.3 *SAN Features*

In this study, the SAN technique (Zhang et al. 2013) was employed to extract both the texture and shape features from optical images in Hong Kong and Sao Paulo, which are shown in Figures 6.15 and 6.16. These SAN feature images show some totally different patterns of the images. In Hong Kong,

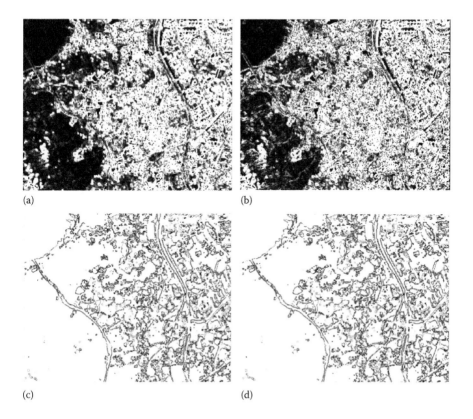

(a)
(b)
(c)
(d)

FIGURE 6.15
SAN-based textures and shape features in Hong Kong. (a) SAN texture (Step = 1), (b) SAN texture (Step = 2), (c) SAN shape feature (Step = 1), and (d) SAN shape feature (Step = 2).

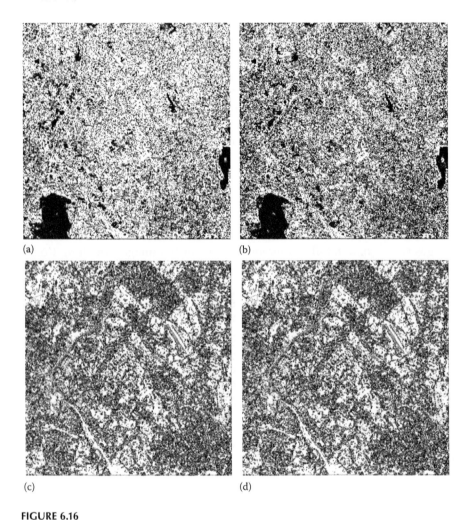

FIGURE 6.16
SAN-based textures and shape features in Sao Paulo. (a) SAN texture (Step = 1), (b) SAN texture (Step = 2), (c) SAN shape feature (Step = 1), and (d) SAN shape feature (Step = 2).

the SAN-based textures were able to separate the sea area from the land and identify the bridge across the sea. Moreover, most of the mountain area could also be separated by the SAN texture images. Residential areas also showed some good consistency in the texture images. For the shape features, which could be obtained by GLCM, the boundaries of buildings, roads, and coastal line could be easily identified (Figure 6.15c and d). As well, the SAN shape features were insensitive to the water quality variation of the sea surface. In addition, both the SAN texture and shape images had much less edge effect compared with the GLCM texture images.

In Sao Paulo, the SAN features also showed very different patterns from the GLCM features (Figure 6.16). However, both advantages and disadvantages could be observed in the SAN feature images in Sao Paulo. Lakes were well separated in the SAN texture features, and the river could be identified from the shape features. Residential areas were highlighted in the shape features, which may be helpful for the classification of impervious surfaces. However, the SAN texture features seemed to add some noises over other land covers with no noticeable separation between different features except water surfaces, which may be negative in the classification results.

6.3.2.2 Features of SAR Images

6.3.2.2.1 GLCM Features

Compared with optical images, SAR images do not carry much information such as rich spectral information. In this study, only single-polarization SAR data was used and thus texture features were calculated from the SAR images. Similarly, six texture measures were employed including MEA, VAR, HOM, DIS, ENT, and SMA to represent the texture features of the SAR images. Even though the SAR data used in this study is TSX data in both Hong Kong and Sao Paulo, they were resampled according to the corresponding optical images. Thus, the final spatial resolutions of TSX images were different in Hong Kong and Sao Paulo. Moreover, the land cover characteristics are also different in the two study cases. Therefore, the GLCM texture features showed a very different view in Hong Kong and Sao Paulo, which can be observed in Figures 6.17 and 6.18.

In Hong Kong, water surfaces on the sea could be easily separated due to their surface geometric characteristics. Tall buildings could be easily identified from the MEA and VAR images due to their high backscattering characteristics. Some man-made objects, such as sets of containers, were highlighted in the DIS image. However, residential area and greening areas in the city could be well separated in the texture images of TSX data, reflected by their similar texture features in Figure 6.17.

In Sao Paulo, the water surfaces and other land covers were well separated in the MEA and VAR features (Figure 6.18). In particular, the whole river could be well identified from the MEA image, which was much better than other features from the optical image. In addition, residential areas with dark impervious surfaces could also be easily observed in the MEA and DIS images. However, it was difficult to separate other land covers, such as bare soil and vegetation, from these TSX feature images. Additionally, the ENT and SMA images generally showed some negative information for all types of land covers, which would add a negative effect to the classification and ISE. Therefore, a feature selection was also needed for the TSX GLCM features in Sao Paulo before classification.

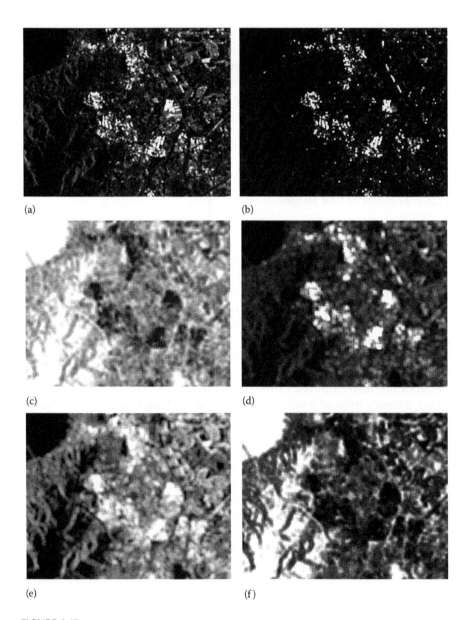

FIGURE 6.17
GLCM-based textures of TerraSAR-X images in Hong Kong. (a) MEA, (b) VAR, (c) HOM, (d) DIS, (e) ENT, and (f) SMA.

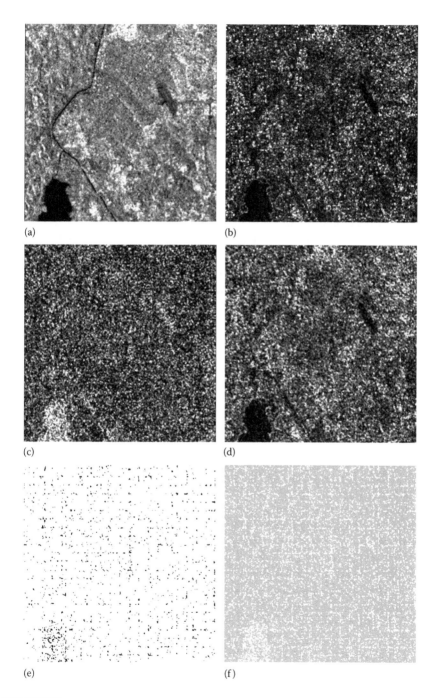

FIGURE 6.18
GLCM-based textures of TerraSAR-X images in Sao Paulo. (a) MEA, (b) VAR, (c) HOM, (d) DIS, (e) ENT, and (f) SMA.

6.3.3 ISE and Comparisons

6.3.3.1 LULC Classification

Figure 6.19 demonstrates the classification results of six land cover types using the combination level of fusion of SPOT-5 and TSX images in Hong Kong. SVM was employed as the fusion and classification method, as SVM has been widely considered to be one of the best and most stable machine-learning approaches. The original SPOT-5 image is provided for a better understanding of the classification results. The result showed that seawater could be well separated using only original optical data, while river water was confused with shaded areas. Moreover, impervious surfaces were mixed with bare soil and vegetation in the central urban areas. The confusion between impervious surfaces and other land covers could be reduced using various feature combinations. However, when using either the GLCM features or the SAN features, edge effects could be observed from Figure 6.19c to f. It shows that both dark and bright impervious surfaces could generally be classified with some small greening areas separated. Shaded areas from tall buildings and hills could also be identified. Fortunately, edge effects were reduced by using the SAN features (Figure 6.19d and f). However, different water qualities in the sea surface lead to some incorrect classification of sea even though the additional use of SAN features and TSX data was able to improve the classification. There were still some parts of the sea surface incorrectly classified as shaded area.

In the classification results of Sao Paulo, different image features demonstrate different impacts on the classification results (Figure 6.20). Using only the original optical image, water surfaces in the rivers and lakes could be well identified. However, some shaded areas from tall buildings were incorrectly classified as water surfaces due to their low reflectance similar to water surfaces. Dark impervious surfaces were confused with shaded areas and vegetation, which produced some noises in the classification results. With the additional use of images features from both optical and SAR data, these confusions could be noticeably reduced, especially the spectral confusion between water surfaces and shaded areas. Noises in the results were also reduced. Nevertheless, edge effects could be observed from all the results using image features, which were caused by the calculation of GLCM texture features using a moving window. With the additional use of SAN features, the confusion between dark impervious surfaces and shaded areas were reduced. However, by using additional SAN features, some shaded areas were incorrectly identified as impervious surfaces, which may cause some negative impacts to the ISE results. Additionally, it should be noted that there are no significant shape features in the optical image in Sao Paulo, mainly due to the medium spatial resolution of Landsat TM. Consequently, the shape features may cause some negative impacts to the results.

To quantitatively assess the classification results, the overall accuracy and Kappa coefficient were calculated and are shown in Table 6.15. In the Hong

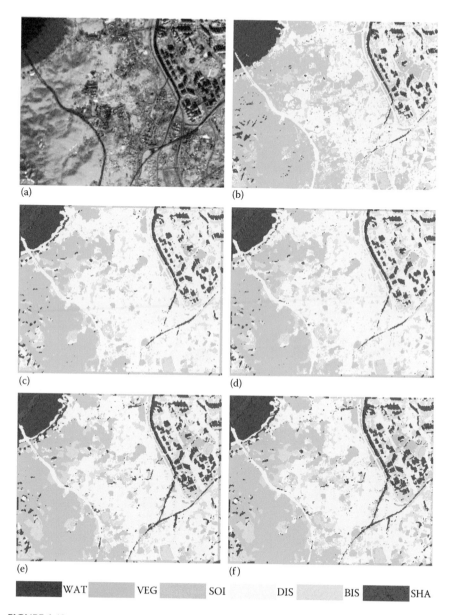

FIGURE 6.19
LULC classification results in Hong Kong. (a) SPOT-5 image (R-G-B = bands 4-2-1), (b) OPT, (c) OPT_SG, (d) OPT_SGS, (e) OPT_SAR, and (f) OPT_SARS. OPT = original optical image alone; OPT_SG = spectral and GLCM features of the optical image; OPT_SGS = spectral, GLCM, and SAN features of the optical image; OPT_SAR = spectral and GLCM features of the optical image as well as intensity and GLCM features of the SAR image; OPT_SARS = spectral, GLCM, and SAN features of the optical image as well as intensity and GLCM features of the SAR image; WAT = water; VEG = vegetation; SOI = bare soil; DIS = dark impervious surfaces; BIS = bright impervious surfaces; SHA = shaded area.

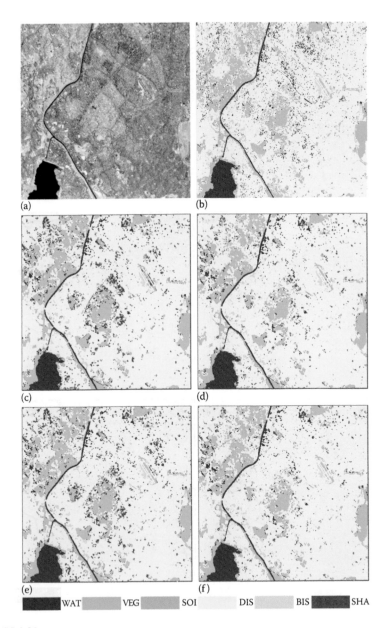

FIGURE 6.20
LULC classification results in Sao Paulo. (a) Optical image, (b) OPT, (c) OPT_SG, (d) OPT_SGS, (e) OPT_SAR, and (f) OPT_SARS. OPT = original optical image alone; OPT_SG = spectral and GLCM features of the optical image; OPT_SGS = spectral, GLCM, and SAN features of the optical image; OPT_SAR = spectral and GLCM features of optical image as well as intensity and GLCM features of the SAR image; OPT_SARS = spectral, GLCM, and SAN features of the optical image as well as intensity and GLCM features of the SAR image; WAT = water; VEG = vegetation; SOI = bare soil; DIS = dark impervious surfaces; BIS = bright impervious surfaces; SHA = shaded area.

TABLE 6.15

Accuracy Assessment of Different Combinations of Datasets and Features

Feature Combination	Hong Kong		Sao Paulo	
	Overall Accuracy	Kappa Coefficient	Overall Accuracy	Kappa Coefficient
OPT	84.05%	0.8086	84.96%	0.8009
OPT_SG	86.76%	0.8409	90.14%	0.8695
OPT_SGS	87.36%	0.8481	89.65%	0.8627
OPT_SAR	88.16%	0.8577	91.03%	0.8813
OPT_SARS	88.77%	0.8649	90.06%	0.8677

Kong case, with only original SPOT data, the overall accuracy was only 84.05% and the Kappa coefficient was 0.8086. When the features of a SPOT image were added, the overall accuracy was 86.76% and the Kappa value was 0.8409 by using the spectral features and GLCM texture features. The overall accuracy increased to 87.36% and Kappa increased to 0.8481 by combining the SAN texture and shape features. By fusing both the SPOT and TSX data, the accuracy was improved to more than 88% for the overall accuracy and more than 0.85 for the Kappa coefficient. Similarly, the additional use of SAN features improved the accuracy during the fusion of SPOT-5 and TSX data. Therefore, the best result came from the use of SPOT-5 and TSX data combined with the use of SAN features of the SPOT-5 image, with an overall accuracy of 88.77% and a Kappa coefficient of 0.8649.

In Sao Paulo, the lowest accuracy was also obtained from using only original Landsat TM data, with an overall accuracy of 84.96% and a Kappa coefficient of 0.8009. Classification accuracy was increased by using image features of Landsat TM image (overall accuracy: 90.14%, Kappa coefficient: 0.8695), and it was further increased by the additional use of TSX data (overall accuracy: 91.03%, Kappa coefficient: 0.8813). However, the additional use of SAN features from optical images did not show positive impacts on the classification. Even though SAN shape features provide some useful information, shape features in medium-resolution images are not significant. In contrast, the shape features in medium-resolution images produce some noises that cause negative impacts to the classification. Consequently, SAN features had a slight negative impact on the final classification results. This result indicates that the effectiveness of SAN features depends on the spatial resolution of the satellite data.

6.3.3.2 ISE

By combining the dark and bright impervious surfaces, and the vegetation, water, soil, and shaded areas, estimations of impervious surfaces were produced as shown in Figures 6.21 and 6.22. From the estimation results in the Hong Kong case (Figure 6.21), it can be seen that some shaded areas (e.g., area

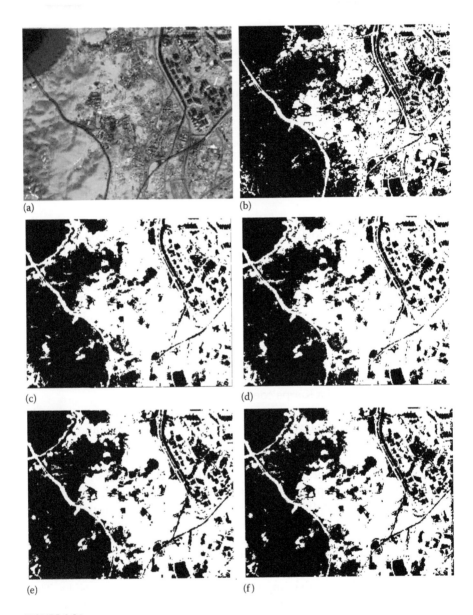

FIGURE 6.21
ISE by combining LULC subtypes in Hong Kong. (a) Optical image, (b) OPT, (c) OPT_SG, (d) OPT_SGS, (e) OPT_SAR, and (f) OPT_SARS. OPT = original optical image alone; OPT_SG = spectral and GLCM features of the optical image; OPT_SGS = spectral, GLCM, and SAN features of the optical image; OPT_SAR = spectral and GLCM features of the optical image as well as intensity and GLCM features of the SAR image; OPT_SARS = spectral, GLCM, and SAN features of the optical image as well as intensity and GLCM features of the SAR image.

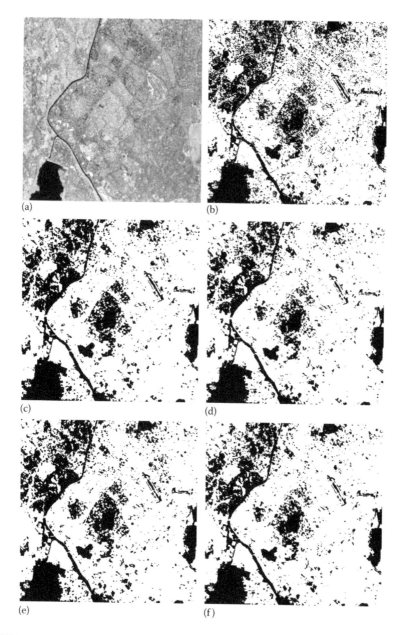

FIGURE 6.22
ISE results in Sao Paulo. (a) Optical image, (b) OPT, (c) OPT_SG, (d) OPT_SGS, (e) OPT_SAR, and (f) OPT_SARS. OPT = original optical image alone; OPT_SG = spectral and GLCM features of the optical image; OPT_SGS = spectral, GLCM, and SAN features of the optical image; OPT_SAR = spectral and GLCM features of the optical image as well as intensity and GLCM features of the SAR image; OPT_SARS = spectral, GLCM, and SAN features of the optical image as well as intensity and GLCM features of the SAR image.

along the sea) and bare soils (located on the mountains) were misclassified as impervious surfaces for all four combinational fusions. In this study, shaded areas were treated as nonimpervious surfaces, and thus confusion between shaded areas and water surfaces on the sea did not impact the result of ISE. Again, SAN features are able to reduce the edge effects of ISE. These types of edge effects are not only located on the edges of the whole image, but are also located on the boundaries between different land objects, which is the main reason why SAN is able to improve the classification accuracy.

The ISE in Sao Paulo demonstrates a consistent result to the LULC classification (Figure 6.22). With only the original Landsat TM image, spectral confusions between dark impervious surfaces and vegetation and between dark impervious surfaces and shaded areas produced misclassifications in the ISE result (Figure 6.22b). After combining the image features from both optical and SAR images, misclassifications such as noises in the residential areas were noticeably reduced. However, similar to the LULC classification, the additional use of image features produced edge effects in the results. Moreover, from the ISE results using additional SAN features (Figure 6.22d and f), an overestimation of impervious surfaces can be observed over the whole area, which was caused by the noisy SAN shape features.

To assess the accuracy of ISE, confusion-matrix-based accuracies were calculated as shown in Table 6.16. After combining different land covers, the general accuracy was increased as the incorrectness between subtypes of impervious surfaces and nonimpervious surfaces was reduced. In Hong Kong, the general pattern of the accuracy was similar to that of the LULC classification. The combined use of SPOT-5 and TSX data increased the accuracy of ISE. The highest accuracy of ISE came from the use of SPOT and TSX combined with the additional SAN features, and the overall accuracy was 97.49% with the Kappa value of 0.9467. However, a slight decrease of accuracy can be observed when combining the SAN with the single use of SPOT data, where the overall accuracy decreased from 96.69% to 96.59, and the Kappa value decreased from 0.9295 to 0.9273. This was caused by the combination of different subtypes of land covers. As discussed in Section 7.5.2, SAN feature

TABLE 6.16

Accuracy Assessment of Different Combinations of Dataset and Features

Feature Combination	Hong Kong		Sao Paulo	
	Overall Accuracy	Kappa Coefficient	Overall Accuracy	Kappa Coefficient
OPT	96.49%	0.9251	91.56%	0.8315
OPT_SG	96.69%	0.9295	91.67%	0.8335
OPT_SGS	96.59%	0.9273	91.35%	0.8270
OPT_SAR	96.89%	0.9337	92.97%	0.8594
OPT_SARS	97.49%	0.9467	92.00%	0.8400

provide better hints to separate the different water qualities of the sea surface as well as to reduce the edge effects, and thus, with SAN feature, fewer
water surfaces were misclassified as shaded areas. However, the confusion
between water surfaces and shaded areas was removed during the estimation of impervious surfaces. Therefore, improvement by using SAN in the
LULC classification will be reduced in the ISE.

In Sao Paulo, the improvement of using various images features can also
be observed from the accuracy assessment (Table 6.16). The lowest accuracy
comes from using only original the optical image, with an overall accuracy of
91.56% and a Kappa coefficient of 0.8315. The overall accuracy was improved
to 91.67% by using additional features of optical images, and to 92.97% by
using features of both optical and SAR images, which was also the highest
accuracy in the Sao Paulo case. The improvement of Kappa coefficient was
0.8335 and 0.8594, respectively. However, the negative effect of SAN features
in the Sao Paulo case is also shown in Table 6.16, indicating that SAN features
are only effective in high-resolution images.

6.3.4 Discussion and Implications

An experiment was designed and conducted to evaluate the efficiency of
different image features of optical and SAR images for ISE. Spectral, texture,
and shape features were extracted from optical and SAR images. Two sets of
optical and SAR images in Hong Kong and Sao Paulo were tested with different spatial resolutions. Both LULC classification and ISE were conducted
separately to investigate the effectiveness of different feature combinations.
Some interesting results were found from the experiments. First, with various
features extracted from the images, accuracy was improved compared with
using only the original image data of optical images. This indicates the effectiveness of feature extraction for remote sensing classifications for LULC and
ISE. Second, with the spectral, texture, and shape feature extraction, the combination of optical and SAR images obtained better results than using optical
data alone. This is consistent to the results in previous experiments, and also
proves the effectiveness of the synergistic use of optical and SAR data. Third,
edge effects located on the image edges of the study areas can be found in
both the study cases due to the texture extraction using GLCM technique as
it applies the moving window with a certain window size. The additional
use of SAN texture and shape features was able to reduce this edge effect to
some extent. When calculating GLCM features, a fixed size of rectangular
neighborhood (moving window) is compulsory, and thus provides the edge
effect on the image boundaries. However, as the size and shape of neighborhood is feasible when calculating SAN features, there is no edge effect on the
image boundaries. Moreover, the SAN shape feature was able to enhance the
edge information on the boundaries between different land objects, which
is the main reason why SAN was able to improve the classification accuracy

in the Hong Kong study case. However, shape features are much more significant in high spatial resolution images (e.g., SPOT-5 images) than low and medium spatial resolution images (e.g., Landsat TM images). In contrast, the use of shape features in low and medium resolution images may bring some noise and consequently cause some negative impacts. This is exactly the situation shown in the Sao Paulo study case.

7

In-Depth Study: ISE Using Optical and SAR Data

7.1 Introduction

Urban impervious surfaces, such as transport-related land (e.g., roads, streets, and parking lots) and building rooftops (commercial, residential, and industrial areas), have been widely recognized as important indicators for urban environments (Arnold and Gibbons 1996; Hurd and Civco 2004; Weng 2001; Weng et al. 2006). Remote sensing has become the major technique to estimate impervious surfaces due to its low cost and convenience for impervious surface mapping on local to global scales. Numerous methods have been proposed to estimate impervious surfaces from remotely sensed images, including subpixel approaches (e.g., the SMA method [Wu and Murray 2003], classification and regression tree model [Yang et al. 2003b], ANN [Weng and Hu 2008], and SVM [Sun et al. 2011]), and per-pixel approaches such as conventional classification methods (Weng 2012). Recently, a BCI was proposed to extract urban impervious surfaces following the VIS conceptual model (Deng and Wu 2012). However, most of these approaches were proposed with optical remote sensing images, and accurate estimation of impervious surfaces remains challenging due to the diversity of urban land covers, leading to difficulties of separating different land covers with similar spectral signatures (Weng 2012). For instance, dry soils or sands are reported to be confused with bright impervious surfaces due to their high reflectance, while water and shade tend to be confused with dark impervious surfaces.

The use of multisatellite images is considered as one promising approach to improve the accuracy of impervious surfaces (Weng 2012). SAR is able to provide useful information about urban areas because it is sensitive to the geometric characteristics of urban land surfaces (Calabresi 1996; Henderson and Xia 1997; Soergel 2010; Tison et al. 2004; Zhang et al. 2012), and thus SAR has been identified as an important source to help extract impervious surfaces with optical data (Jiang et al. 2009; Weng 2012; Yang et al. 2009a). Fusion between optical and SAR data can be performed on three different levels: the pixel level, feature level, and decision level. Pixel-level fusion is reported as inappropriate for SAR images because of speckle noises (Zhang et al. 2010). For feature-level fusion, several approaches have been proposed including layer-stacking and ensemble-learning methods (e.g., bagging, boosting, AdaBoost, and RF [Hall and Llinas 1997; Rokach 2010]). The ensemble-learning methods can be combined with different classifiers (e.g., ANN and SVM [Rokach 2010]). For decision-level fusion, various weighting methods (e.g., majority voting, entropy weighting, and performance weighting) and the Dempster-Shafer theory have been applied. However, conventional classifiers with a layer-stacking technique are not appropriate in this case as optical reflectance and SAR backscattering data do not correlate (Zhang et al. 2010). Among these methods, the DT method will be given more attention, while RF has been reported to perform extremely well in the fusion of optical and SAR data (Waske and van der Linden 2008). However, the potential and effectiveness of RF on the fusion between optical and SAR images needs to be explored, especially in terms of the estimation of urban impervious surfaces.

This chapter aims to evaluate the effectiveness of RF to synergistically combine the optical and SAR data in terms of ISE. A combination of pixel-level and feature-level fusion methods is adopted. Additionally, the Kappa coefficient based on the confusion matrix and OOB error built into the RF are compared to assess the effectiveness of fusing optical and SAR images.

7.2 Study Areas and Datasets

7.2.1 Study Areas

Two groups of study areas were employed for this comprehensive study. One is used to investigate the optimal parameters of the RF algorithm for ISE, and the other was used to test the effectiveness of RF for ISE by comparatively using two different sets of SAR images.

First, three cities, Guangzhou, Shenzhen, and Hong Kong, located in the PRD were selected as the study areas to investigate the optimization of the RF algorithm. Detailed introduction to the environment and socio-economic background of these three cities can be found in Section 3.1.1.

Second, in order to evaluate the potential of RF algorithm for ISE comprehensively, three cities, Shenzhen, Mumbai and Sao Paulo, located in different regions of the tropical and subtropical areas, were employed as the study areas. As described in Chapter 3:

1. Shenzhen is located in a subtropical humid climate region in the Southern hemisphere of the earth.
2. Mumbai is located in a tropical wet and dry climate region and has been undergoing dramatic urbanization process. However, there are

many problems introduced by the rapid urbanization, such as urban fragmentation (Gandy 2008). Therefore, remote sensing of the urbanization process of Mumbai would be a good way to monitor the urban sprawl in order to improve urban planning and management.

3. Sao Paulo is another subtropical city in Brazil with a subtropical humid climate. The Sao Paulo metropolitan area has been undergoing a rapid urbanization process since the twenthieth century. However, urbanization has brought significant environmental impacts to the ecosystem such as the deforestation of the rainforest (Torres et al. 2007). Therefore, estimation of impervious surfaces would be beneficial for urban planning and environmental management of the city.

7.2.2 Satellite Data and Coregistration

For the first group of study areas, three different combinations of optical and SAR satellite data sets were selected for the three cities (Table 7.1). For Guangzhou, a scene of a Landsat ETM+ image and a scene of an ENVISAT ASAR WSM image were employed. The ENVISAT ASAR WSM data was obtained on the descending direction with V/V polarization and a pixel size of 75 m. For Shenzhen, a scene of a SPOT-5 image and a scene of ENVISAT ASAR ASA_IMP_1P data were used. The spatial resolution of the ASAR IMP data was 12.5 m. For Hong Kong, a SPOT-5 and a SAR image from TerraSAR-X were employed. The TSX image used in this study was obtained in the StripMap mode with a spatial resolution of 3 m.

The second group of datasets included two types of SAR images for each study site (Table 7.2). For Shenzhen, a scene of a SPOT-5 image, a scene of

TABLE 7.1

Datasets of the First Group for Optimization Investigation of the RF Algorithm

Study Site	Optical Image	SAR Image
Guangzhou	Landsat ETM+	ENVISAT ASAR
Shenzhen	SPOT-5	ENVISAT ASAR
Hong Kong	SPOT-5	ENVISAT ASAR

TABLE 7.2

Datasets of the Second Group for ISE Using Optimized RF

Study Site	Optical Image	SAR Image	SAR Image
Shenzhen	SPOT-5	ENVISAT ASAR	TerraSAR-X
Mumbai	Landsat TM	ENVISAT ASAR	TerraSAR-X
Sao Paulo	Landsat TM	ENVISAT ASAR	TerraSAR-X

TABLE 7.3

Coregistration Design between Optical and SAR Images

| Optical Image (Base Image) | | SAR Image | | Registered Result |
Satellite Sensor	Resolution (*m*)	Satellite Sensor	Resolution (*m*)	Resolution (*m*)
Landsat TM/ETM+	30	ENVISAT ASAR (WSM)	75	30
Landsat TM/ETM+	30	ENVISAT ASAR (IMP)	12.5	30
Landsat TM/ETM+	30	TerraSAR-X	3	30
SPOT-5	10	ENVISAT ASAR (IMP)	12.5	10
SPOT-5	10	TerraSAR-X	3	10

ENVISAT ASAR ASA_IMP_1P data, and a scene of TerraSAR-X data were used. The SPOT-5 image consisted of four bands with 10 m resolution. The ENVISAT ASAR image was of V/V polarization with 12.5 m resolution, while the TerraSAR-X image was in the StripMap mode with a spatial resolution of 3 m. Datasets for Mumbai and Sao Paulo were of the same category. For optical data, Landsat TM images with 30 m resolution were used. For SAR images, both ENVISAT ASAR and TerraSAR-X images were used. The ENVISAT ASAR data was obtained in IMP mode with 12.5 m resolution, while the TerraSAR-X data was acquired in StripMap mode with 3 m resolution.

After preprocessing all the satellite images, both the optical images and SAR images were coregistered to the same georeference system of the UTM projection (Zone 50 N) and Datum WGS84. Over 20 control points were manually selected for each pair of optical and SAR images, and the linear transformation approach wass used to conduct the coregistration. The spatial resolutions of the final registered images were determined by the corresponding optical image, which was clearer for human visual interpretation and thus easier for the manual selection of control points (Table 7.3). The RMS error (RMSE) of the coregistration for each pair of optical and SAR data was less than half a pixel.

7.3 Feature Extraction of Optical and SAR Data

The traditional feature extraction methods introduced in Section 3.5.3.1 were employed in this chapter to extraction features from both optical and SAR images. If we combine the SAR image as a band to the optical images, then

TABLE 7.4

Number of Bands and Features for the First Group of Datasets

Study Area	Number of Bands	Number of Texture Feature Images	Total Number of Images Input into RF
Guangzhou	7	28	35
Shenzhen	5	20	25
Hong Kong	5	20	25

TABLE 7.5

Number of Features for the Second Group of Datasets

Study Area	Optical	Optical + ASAR	Optical + TSX
Shenzhen	4	33	33
Mumbai	6	49	49
Sao Paulo	6	49	49

we can calculate the total number of images (bands) that are the input of RF. Table 7.4 shows the number of bands and number of texture features from the first group of datasets, and the total number of images input into the RF algorithm. It shows there are 35 variables in total for the Guangzhou case, 25 variables for Shenzhen, and 25 variables for Hong Kong.

For the second group of datasets, the number of features was also calculated after applying GLCM texture feature extraction to the optical and SAR images. Table 7.5 shows the number of features in different combinations of optical and SAR datasets. First, when using only optical images, there are only four features (the original multispectral bands) for the Shenzhen data, six features for Mumbai, and six features for Sao Paulo. Second, when combined with SAR data (ASAR or TSX), there were 33 features for Shenzhen and 49 features for Mumbai and Sao Paulo.

7.4 Classification Strategy and Accuracy Assessment

Impervious surface mapping at the per-pixel level is actually a classification task, where impervious and nonimpervious surfaces are a combination of various land cover types. Conventional LULC includes vegetation, urban areas, and water, and each land cover type shares similar spectral and spatial characteristics. Therefore, they are often identified individually during the classification procedure. However, impervious/nonimpervious surfaces consist of various land cover materials. For instance, impervious surfaces can be made up of dark material (e.g., asphalt and old concrete) and bright material (e.g., new concrete and metal), while nonimpervious surfaces are also made up of very diverse material (e.g., vegetation, water, and base soils). In this study, a two-step approach was employed to estimate impervious surfaces. First, six land cover types—dark impervious surfaces, bright impervious surfaces, vegetation, water body, bare soil, and shaded areas—were identified with a classification procedure using RF. Second, a combination procedure was conducted to combine various land covers into impervious and nonimpervious surfaces.

In particular, shaded areas are treated as a single land cover type as they often have unique spectral and spatial characteristics. Moreover, since shaded areas may be impervious (e.g., roads and rooftops) or nonimpervious (e.g., greening areas), they are treated as nonimpervious surfaces in the second step of combination in this study. Therefore, dark impervious surfaces and bright impervious surfaces are combined as impervious surfaces, and vegetation, water, bare soil, and shade are combined as nonimpervious surfaces. Additionally, as misclassification may happen not only between impervious and nonimpervious land cover types, but also among different subtypes of impervious or nonimpervious types, the accuracy of classification before and after the combination operation may be different. Therefore, in this study, accuracy assessment is conducted on the classification results before and after the combination of impervious/ nonimpervious surfaces.

Two accuracy indices are employed to assess the accuracy of ISE. One is the OOB error, which is built into the RF algorithm. OOB error is calculated based on the training samples, which are separated into two parts in the RF algorithm; one part for constructing the RF, and the other for evaluating the performance of RF. However, the lowest OOB does not necessarily guarantee the best performance of a RF when it is applied to datasets other than the training samples. Therefore, the overall accuracy and Kappa coefficient based on the confusion matrix are also employed to assess the accuracy (Jensen 2007). In addition, reference data is collected through visual interpretation of the optical and SAR images in the three study areas. Higher-resolution satellite images from Google Earth near the corresponding dates are used to help the visual interpretation. Moreover, Orthophoto in Hong Kong, with 0.5×0.5 m resolution, was purchased from the Hong Kong governmental agency to help improve the quality of visual interpretation of the Hong Kong images. Lastly, 1528 samples were collected for the Guangzhou area, 1949 samples were collected for the Shenzhen area, 1537 samples were collected for the Hong Kong area, 1721 samples were collected for the Mumbai area, and 1680 samples were collected for the Sao Paulo area. Of these reference samples, 50% were used as the training samples to construct the RF and 50% were used as the test samples to validate the results and test the effectiveness of the method.

7.5 Optimization of RF

7.5.1 Determining the Optimal Number of Features in Each Decision Tree

To test the impact of the number of variables selected for splitting each node in the decision trees, the number was changed from 1 to the total number of variables. Meanwhile, as the number of decision trees in the RF also influences the results, four levels of the number of decision trees were selected to test this influence; that is, 5, 10, 15, and 20, respectively. The Kappa coefficient and the OOB error built into the RF were employed to assess the accuracy of ISE. Figure 7.1 illustrates the influences of the variation of the number of variables under the selected four different numbers of decision trees.

First, for the Kappa coefficient, a similar pattern was observed for all four different numbers of decision trees. The Kappa coefficient increased quickly at first as the number of selected variables increased, reached a peak, and then decreased steadily with slight fluctuation. The peaks of the curves are located on different positions for the three study areas. They are approximately located on eight variables, six variables, and six variables for Guangzhou, Shenzhen, and Hong Kong, respectively. In addition, the different numbers of decision trees can have significant impact on the Kappa coefficient. The results demonstrate that more decision trees tend to produce a more accurate result, as shown in Figure 7.1 (a, c, and e). However, the gap between two neighboring curves became smaller and smaller as the number of decision trees increased, and there was a large area of overlay between the two curves for 15 and 20 decision trees in all three study cases.

Second, the OOB error varied with the number of decision trees and study cases, as shown in Figure 7.1 (b, d, and f). However, the changing pattern of the OOB error was different from that of the Kappa coefficient. At the beginning stage, the OOB error decreased quickly as the number of selected variables increased and reached the lowest point. From then on, an increase of the number of variables did not have significant impact on the OOB error, which became relatively steady. Therefore, the curves were almost parallel in the last part, although the number of decision trees differed. For different study cases, the OOB error first reached its lowest point when the number of selected variables was 9, 7, and 6, for Guangzhou, Shenzhen, and Hong Kong, respectively. However, the number of decision trees had significant impact on the accuracy for a given case even though the pattern of each curve was similar. More decision trees tended to produce lower OOB curve steady values. However, the gaps between two neighboring curves also became narrower and narrower when the number of decision trees increased.

As reported in previous research, the number of features in each decision tree was suggested to be the root of the total number of variables from an empirical point of view (Gislason et al. 2006; Stumpf and Kerle 2011).

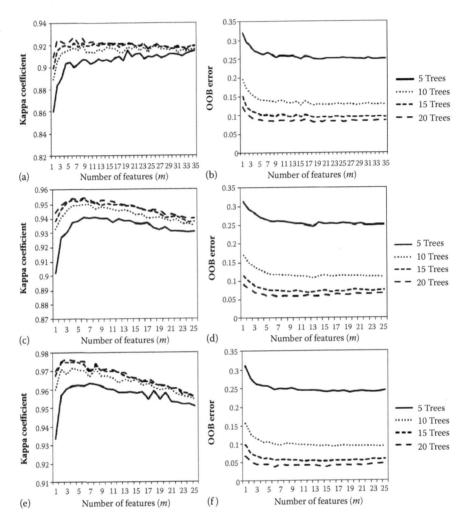

FIGURE 7.1
Impacts of different numbers of variables (or features) in each decision tree. (a) Guangzhou: Kappa coefficient, (b) Guangzhou: OOB error, (c) Shenzhen: Kappa coefficient, (d) Shenzhen: OOB error, (e) Hong Kong: Kappa coefficient, and (f) Hong Kong: OOB error.

Thus, according to Gislason et al. (2006) and Stumpf and Kerle (2011), the optimal number of features in each decision tree should be 6, 5, and 5 for Guangzhou, Shenzhen, and Hong Kong, as there are 35, 25, and 25 variables in total, respectively (Table 7.4). In this study, the observed optimal number of features was 8, 6, and 6, considering the Kappa coefficient, while this value was 9, 7, and 6, considering the OOB error. This result indicates that the optimal number of variables should be a little bit higher than the root of the total number of variables. In addition, the optimal numbers of variables are not exactly the same by considering the best Kappa coefficient and by considering the lowest OOB error, even though they are closed. This result also indicates that only the built-in accuracy assessment of RF (OOB error) may not reflect the real accuracy of the classification result, as the optimal number of variables differs based on the Kappa coefficient and OOB error. Therefore, additional testing data is needed in order to evaluate the accuracy of the classification using RF. In this study, we set the optimal number of variables according to the best Kappa coefficient, since additional testing samples are more frequently used to validate the classification results in remote sensing applications. Moreover, in order to provide a reference for further similar applications, a simple rule of determining the optimal number of variables in RF is given in Equation 7.1 according to the experiment results and discussion.

$$m = \lfloor \sqrt{M} \rfloor + 1 \tag{7.1}$$

where m is the optimal number of variables to determine the nodes in a decision tree in RF and M is the total number of variables. The function $\lfloor x \rfloor$ is the largest integer not greater than x. Equation 7.1 indicates that the optimal number of variables is a little bigger than the root of the total number of variables.

7.5.2 Determining the Optimal Numbers of Decision Trees in the RF

According to the results of Figure 7.2, the impacts of the number of decision trees on the classification accuracy can be significant. Even though this impact tends to be reduced when the number of decision trees increases, there are only four different numbers of decision trees tested, and further experiments are needed in order to gain insight into the impact of this factor. In this experiment, the number of decision trees was changed to a larger range from 1 to 60. Similarly, four different numbers of variables were selected. However, since the total numbers of variables were different for every study cases, 3, 6, 9, and 12 variables were selected for Guangzhou, while 2, 5, 8, and 11 variables were used for Shenzhen and Hong Kong. Corresponding results are illustrated in Figure 7.2, where some interesting findings are demonstrated.

First, the Kappa coefficient showed a very tight and similar pattern for the four different numbers of variables; that is, it increased quickly as the number of decision trees went up and then reached a maximal number. From then on, the Kappa coefficient was relatively steady when the number of decision trees continued to increase. What is more interesting is that even though the selected number of variables, the sensors, and the spatial resolutions were different in the three study cases, the starting points where the Kappa coefficient became steady are almost the same, which is approximately 20 decision trees in this research. Nonetheless, the maximal Kappa coefficients are

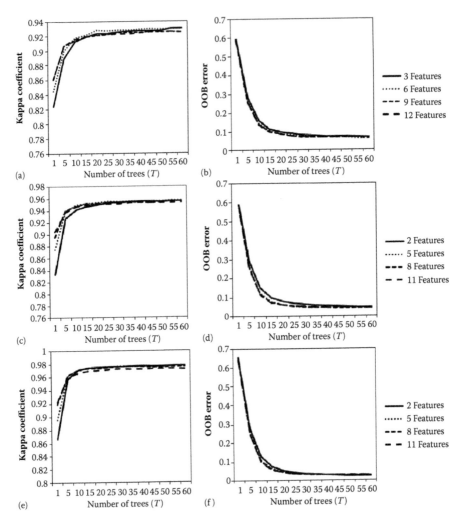

FIGURE 7.2
Impacts of different numbers of trees in the RF. (a) Guangzhou: Kappa coefficient, (b) Guangzhou: OOB error, (c) Shenzhen: Kappa coefficient, (d) Shenzhen: OOB error, (e) Hong Kong: Kappa coefficient, and (f) Hong Kong: OOB error.

different for the three cases, which were about 0.92 for Guangzhou, 0.95 for Shenzhen, and 0.97 for Hong Kong.

Second, the RF built-in accuracy, the OOB error, demonstrated a consistent result with the variation of selected number of variables and three study cases. For all four different numbers of variables, the changing pattern of the curves was very similar and the gap between two curves was extremely close. The OOB error first dropped down rapidly and then became steady after the number of decision trees reached 20, which is consistent with the changing pattern of the Kappa coefficient. Additionally, the steady values of the OOB error were different in the three study cases. The minimal value of OOB error was approximately 0.09 for Guangzhou, 0.08 for Shenzhen, and 0.05 for Hong Kong.

The experiment results indicate that the optimal number decision trees is independent of the number of selected variables for splitting each node in a decision tree, and it is also independent of the types of sensors and the spatial resolutions of remote sensing images. Since the construction of more decision trees requires more building time of the RF, the optimal number of decision trees should be the first priority when the Kappa coefficient reaches its highest point and the OOB error reaches its lowest point. In specific application, this optimal number of decision trees can be determined by a statistical procedure similar to this experiment, and for this study, 20 decision trees are best for a RF to combine the optical and SAR data for ISE.

7.6 ISE with Optimized Models

In order to perform RF to classify impervious surfaces using both optical and SAR images, the optimal parameters should be used to configure the RF. According to the conclusion of Section 7.5, the optimal number of features (m) and the optimal number of decision trees (T) are shown in Table 7.6. Notice that the optimal number of features was calculated using Equation 7.1, while the optimal number of decision trees was set as 20 according to the experimental results.

Therefore, impervious surfaces were classified using the second group of datasets with the optimal parameters in Table 7.6. In order to provide a better understanding of the behaviors of the RF algorithm to classify different land covers types, the LULC result before combining the impervious and nonimpervious land covers is provided, with the detailed confusion matrices shown in Tables 7.7 and 7.8. The comparison of ISE using single optical data and combined optical and ASAR data is demonstrated in Table 7.7, while the comparison of ISE using single optical data and combined optical and TSX data is shown in Table 7.8. The result of using single optical data is listed in

TABLE 7.6

Parameter Settings for the Number of Features (m) and Decision Trees (T)

	Optical		Optical + ASAR		Optical + TSX	
Study Area	m	T	m	T	m	T
Shenzhen	3	20	6	20	6	20
Mumbai	3	20	8	20	8	20
Sao Paulo	3	20	8	20	8	20

TABLE 7.7

Confusion Matrices for Urban Land Cover Classification (ASAR)

	Optical						Optical + ASAR					
	VEG	DIS	BIS	WAT	SOI	SHA	VEG	DIS	BIS	WAT	SOI	SHA
Shenzhen												
VEG	128	2	0	4	0	1	131	1	0	3	0	0
DIS	2	95	17	0	2	11	0	119	0	0	0	8
BIS	0	8	104	0	2	0	0	0	114	0	0	0
WAT	3	0	0	91	0	4	0	0	0	96	0	2
SOI	0	4	8	0	102	0	0	0	1	0	113	0
SHA	2	18	0	2	0	101	1	8	0	0	0	114
	OA: 87.34% Kappa: 0.8478						OA: 96.62% Kappa: 0.9594					
Mumbai												
VEG	177	12	4	0	2	0	185	9	1	0	0	0
DIS	21	188	10	0	4	0	8	211	4	0	0	0
BIS	2	26	125	0	3	0	0	8	148	0	0	0
WAT	1	0	0	148	0	0	0	0	0	149	0	0
SOI	3	9	12	0	113	0	5	4	2	0	126	0
SHA	0	0	0	0	0	0	0	0	0	0	0	0
	OA: 87.33% Kappa: 0.8398						OA: 95.89% Kappa: 0.9507					
Sao Paulo												
VEG	244	2	0	0	1	0	247	0	0	0	0	0
DIS	11	244	2	2	6	0	3	254	1	0	7	0
BIS	2	8	126	0	7	0	0	1	140	0	2	0
WAT	0	0	0	141	0	0	0	0	0	141	0	0
SOI	1	9	0	0	34	0	1	7	1	0	35	0
SHA	0	0	0	0	0	0	0	0	0	0	0	0
	OA: 93.93% Kappa: 0.9194						OA: 97.26% Kappa: 0.9637					

Note: BIS = bright impervious surfaces; DIS = dark impervious surfaces; OA = overall accuracy; SHA = shaded area; SOI = soil; VEG = vegetation; WAT = water.

both Tables 7.7 and 7.8 to provide a better comparison of the effectiveness of using different SAR images.

Several important findings can be observed in Table 7.7 by combining optical and ASAR images. First, in the Shenzhen case, dark impervious surfaces (DIS) were easily confused with shade (SHA) when using only an optical image. For instance, 18 pixels of SHA were mistakenly classified as DIS, while 11 pixels of DIS were classified as SHA. Therefore, there are 29 pixels in total that are incorrectly classified. Moreover, bright impervious surfaces (BIS) and DIS were also confused with each other. However, after combining the optical and ASAR images, these incorrect pixels were correctly classified. For instance, only eight pixels of SHA were classified as DIS and eight DIS pixels were classified as SHA. The total number misclassified pixels for

TABLE 7.8

Confusion Matrices for Urban Land Cover Classification (TSX)

	Optical						Optical + TSX					
	VEG	DIS	BIS	WAT	SOI	SHA	VEG	DIS	BIS	WAT	SOI	SHA
Shenzhen												
VEG	128	2	0	4	0	1	132	1	0	2	0	0
DIS	2	95	17	0	2	11	0	120	0	0	0	7
BIS	0	8	104	0	2	0	0	0	114	0	0	0
WAT	3	0	0	91	0	4	0	0	0	98	0	0
SOI	0	4	8	0	102	0	0	0	0	0	114	0
SHA	2	18	0	2	0	101	1	9	0	0	0	113
	OA: 87.34% Kappa: 0.8478						OA: 97.19% Kappa: 0.9662					
Mumbai												
VEG	177	12	4	0	2	0	187	8	0	0	0	0
DIS	21	188	10	0	4	0	11	202	9	0	1	0
BIS	2	26	125	0	3	0	3	8	145	0	0	0
WAT	1	0	0	148	0	0	1	0	0	148	0	0
SOI	3	9	12	0	113	0	4	4	3	0	126	0
SHA	0	0	0	0	0	0	0	0	0	0	0	0
	OA: 87.33% Kappa: 0.8398						OA: 93.95% Kappa: 0.9237					
Sao Paulo												
VEG	244	2	0	0	1	0	245	0	0	0	2	0
DIS	11	244	2	2	6	0	1	257	0	0	7	0
BIS	2	8	126	0	7	0	0	5	135	0	3	0
WAT	0	0	0	141	0	0	0	0	0	141	0	0
SOI	1	9	0	0	34	0	1	6	2	0	35	0
SHA	0	0	0	0	0	0	0	0	0	0	0	0
	OA: 93.93% Kappa: 0.9194						OA: 96.79% Kappa: 0.9574					

Note: BIS = bright impervious surfaces; DIS = dark impervious surfaces; OA = overall accuracy; SHA = shaded area; SOI = soil; VEG = vegetation; WAT = water.

these two classes was reduced to 16. The confusion between DIS and BIS was noticeably reduced with no incorrectly classified pixels. As a result, with the additional use of an ASAR image, the overall accuracy (OA) was improved from 87.34% to 96.62%, while the Kappa coefficient increased from 0.8478 to 0.9594. Second, in the Mumbai case, DIS seemed to be more easily confused with vegetation (VEG), while BIS was easily confused with bare soil (SOI). Before combing optical and ASAR data, 12 pixels of VEG were classified as DIS and 21 pixels of DIS were classified as VEG. In addition, 12 pixels of BIS were mistakenly classified as BIS. The confusion between DIS and BIS can also be easily observed in Table 7.7 with 36 (= 26 + 10) incorrect classified pixels. These mistakes were dramatically reduced after combining the optical and ASAR data, with only several misclassified pixels between these classes. For instance, nine pixels of VEG were misclassified as DIS and eight pixels of DIS were misclassified as VEG. The misclassification between DIS and BIS was reduced from 36 to 12 pixels. Confusion between SOI and BIS was also greatly reduced with only five misclassified pixels. In general, OA increased from 87.33% to 95.89% and the Kappa coefficient increased from 0.8398 to 0.9507. Third, in the Sao Paulo case, the easily confused classes are between VEG and DIS, between SOI and DIS, and between SHA and the two impervious surface classes. While using optical data alone, 11 DIS pixels were classified as VEG and nine SOI pixels were classified as DIS, 13 (= 6 + 7) pixels of impervious surfaces were incorrectly classified as SHA. After combining optical and ASAR images, only three VEG pixels and seven SOI pixels were incorrectly classified. As a result, OA increased from 93.93% to 97.26% together with the Kappa coefficient, which increased from 0.9194 to 0.9637.

When using TSX data instead of ASAR data to be combined with optical images, different characteristics could be found, as seen in Table 7.8. In Shenzhen, the improvement of using TSX data is noticeable. Only nine pixels of SHA were misclassified as DIS, seven pixels of DIS as SHA, two pixels of VEG as WAT, one pixel of VEG as DIS, and one pixel of SHA as VEG. The OA increased from 87.34% to 97.19%, which is higher than that of combining optical and ASAR data (96.62%). The Kappa coefficient improved from 0.8378 to 0.9662, higher than using ASAR data (0.9594). In Mumbai, improvement could also be observed using TSX data. Misclassification between DIS and VEG was reduced from 33 (= 21 + 12) pixels to 19 (= 11 + 8) pixels. The confusion between DIS and BIS was also reduced. Nevertheless, this improvement was less than that of using ASAR data. Using TSX data, the OA was improved from 87.33% to 93.95%, which was 95.89% using ASAR data. The Kappa coefficient increased from 0.8398 to 0.9237, which was 0.9507 using ASAR data. In Sao Paulo, after combining optical and TSX images, only one DIS pixel and six SOI pixels were incorrectly classified to VEG and DIS, respectively. The confusion between DIS and BIS was reduced from eight pixels to five pixels. As a result, the OA increased from 93.93% to 96.79%, lower than that of using ASAR data (97.26%). The Kappa coefficient increased from 0.9194 to 0.9574, which was lower than that of using ASAR data (0.9637).

In general, several important conclusions can be drawn from the above results. First, when using different combination of optical and SAR images, the land cover classes that are easily confused are not necessarily the same. This may be caused by the types of sensors, the spatial resolutions of the optical and SAR images, and climate of the study areas. For instance, for Mumbai, which is located in a tropical area, the confusion between impervious surfaces and other classes are higher than in other cities. Second, even though the easily confused classes are not always the same in different study cases, the effectiveness of combining the optical and SAR images can be verified and confirmed with an increase of both overall accuracy and

Kappa coefficient. Third, even though TSX data has a much higher spatial resolution than ASAR data, the improvement of using TSX data may not be necessarily higher than using ASAR data. In contrast, combining optical and ASAR data tends to obtain better results, such as in the Mumbai and Sao Paulo cases.

Then, in the second step, DIS and BIS were combined as impervious surfaces (IS), while VEG, WAT, SOI, and SHA were combined as nonimpervious surfaces (NIS). To better understand the results quantitatively, new confusion matrices are computed in Table 7.9. The results in Table 7.9 are generally consistent with those in Tables 7.7 and 7.8, while the OA and Kappa coefficient are generally higher, since the confusion between two impervious classes or two nonimpervious classes is removed after the second step of combination.

Some interesting results can be found in Table 7.9. First, it demonstrates that the misclassification between IS and NIS was dramatically reduced by combining the optical and SAR images for Shenzhen, Mumbai, and Sao Paulo, using either ASAR or TSX images. Second, with either ASAR or TSX data, combining optical and SAR data improved the ISE results compared with using single optical data, which can be observed from the OA and Kappa coefficients in the three study cases. Third, even though combining optical and TSX data yielded less improvement than combining optical and ASAR data in the LULC classification result, this does not guarantee better results after combining the subtypes of IS and NIS. For instance, in Sao Paulo, the

TABLE 7.9

Confusion Matrices for Impervious Surface Mapping

	Optical		Optical + ASAR		Optical + TSX	
	IS	NIS	IS	NIS	IS	NIS
Shenzhen						
IS	224	17	233	8	234	7
NIS	32	438	10	460	10	460
	OA: 93.11%		OA: 97.47%		OA: 97.61%	
	Kappa: 0.8485		Kappa: 0.9436		Kappa: 0.9468	
Mumbai						
IS	349	30	371	8	364	15
NIS	37	444	16	465	15	466
	OA: 92.21%		OA: 97.21%		OA: 96.51%	
	Kappa: 0.8423		Kappa: 0.9435		Kappa: 0.9292	
Sao Paulo						
IS	380	28	396	12	397	11
NIS	11	421	8	424	8	424
	OA: 95.36%		OA: 97.62%		OA: 97.74%	
	Kappa: 0.9070		Kappa: 0.9523		Kappa: 0.9547	

Note: IS = impervious surfaces; NIS = nonimpervious surfaces; OA = overall accuracy.

OA was improved to 97.74% from 95.36% with the additional use of TSX data, while the Kappa value increased to 0.9547 from 0.9070. This improvement is higher than that of using ASAR data. However, in the case of Mumbai, improvement by using TSX data was still not as high as by using ASAR data before and after combining IS and NIS classes.

Generally, backscattering information in SAR imagery can contribute to improving the accuracy of ISE in three different ways. First, since microwaves are very sensitive to the geometric configurations of land surfaces, the backscattering of microwaves carries much information about the geometric features in urban areas, such as the surface roughness determined by buildings and transportation networks. Therefore, SAR images add more distinguishable information between impervious surfaces and nonimpervious surfaces. Second, microwaves are also sensitive to moisture, including the moisture in bare soil and the water content in vegetation. This characteristic makes it easier to separate bare soils and bright impervious surfaces in SAR imagery by reducing the spectral confusion using optical images alone, which is the situation in this study area. Third, SAR remote sensing often works in a side-looking way, leading to a different view angle from that of optical remote sensing. As a result, shaded areas in optical images are often not shaded in the corresponding SAR images, and thus land surface information under the shade in optical images can be seen in SAR images. Therefore, the spectral confusion between shaded areas and dark impervious surfaces in optical images can be significantly reduced with the additional use of SAR images.

7.7 Discussion and Implications

Impervious surfaces are attracting increasing attention because they are not only significant in the urban environment, but also an indicator of the urbanization. Nevertheless, accurate mapping of urban impervious surfaces remains challenging due to their spectral diversity. This chapter presented our efforts to synergistically combine the two data sources to improve the mapping of impervious surfaces using the RF algorithm. Four combinations of optical and SAR images, Landsat TM/ETM+ and ENVISAT ASAR, Landsat TM/ETM+ and TerraSAR-X, SPOT-5 and ENVISAR ASAR, and SPOT-5 and TerraSAR-X, were selected in various study areas including Guangzhou, Shenzhen, Hong Kong, Mumbai, and Sao Paulo to validate the effectiveness of the methods in this study.

Results indicate some interesting findings about the application of RF to the fusion of optical and SAR data. First, the built-in OOB error is insufficient for accuracy assessment, and assessment with additional reference data is required for combining optical and SAR images using RF. In this study, the OA and Kappa coefficient were employed as an additional assessment. The OA and Kappa values shown a consistent pattern with only a slight difference. Second, the optimal number of variables (m) for splitting the decision tree nodes in RF should be somewhat different from the previously reported principle, which indicates m as the root number of the total variables. In this study, an empirical relationship (Equation 7.1) was provided for determining the parameter m. Third, the optimal number of decision trees (T) in RF is not sensitive to the resolutions and sensor types of optical and SAR images, and the optimal T in this study is 20. Fourth, the combined use of optical and SAR images using RF is effective to improve land cover classification and ISE by reducing the confusions between bright impervious surfaces and bare soil, dark impervious surfaces and bare soil, as well as shaded areas and water surfaces. Fifth, two SAR datasets, ASAR and TSX, were comparatively employed in this study, with interesting results indicating that higher-resolution SAR data does not guarantee greater improvement compared to lower-resolution SAR data. Moreover, the effectiveness of various-resolution SAR data may also depend on the classification modes such as the LULC classification and impervious surface mapping. Lastly, even though the easily confused land classes tend to be different in different image resolutions, the effectiveness of combining optical and SAR images is consistent. This improvement is more noticeable for the fusion of optical and SAR images with lower resolutions. The conclusions of this study should serve as an important reference for further applications of optical and SAR images, and as a potential reference for the applications of RF to the fusion of other multisource remote sensing data.

8

Conclusions and Recommendations

Impervious surfaces have been widely recognized as an important land surface component due to their significance in both urban environmental and socioeconomic studies. Therefore, the estimation and mapping of impervious surfaces have become increasingly important all over the world. However, accurate estimation of impervious surfaces is still challenging due to the diversity of urban land covers, which produce various spectral confusions between different land surface materials. Moreover, most of the previous research focused on urbanized areas in temperate latitude regions where many important cities and metropolitans are located. Consequently, the accurate estimation of impervious surfaces in tropical and subtropical regions, where the land cover diversity is unique due to unique seasonal climatology and plant phenology, becomes even more challenging than in other regions of the world. The main objective of this book has been to promote the combined use of optical and SAR images to improve the accuracy of ISE in tropical and subtropical regions. The seasonal characteristics of land covers and their impact on ISE in tropical and subtropical regions have been investigated. This section summarizes the major findings and conclusions of this research, highlights the limitations of the study, and suggests possible research topics for future research.

8.1 Conclusions

8.1.1 Seasonal Effects of ISE in Tropical and Subtropical Areas

Accurate ISE remains challenging due to the diversity of impervious surfaces, and seasonal effects from the climate zones is one of the key issues that influences the accurate estimation of impervious surfaces. In this study, four scenes of Landsat TM/ETM+ images were carefully chosen for four different seasons in four typical cities, Guangzhou, Mumbai, Sao Paulo, and Cape Town, from tropical and subtropical areas, and two classification methods, ANN and SVM, were employed to extract the impervious surfaces from the images at the pixel level. The experimental results demonstrate quite a unique view of seasonal effects in tropical and subtropical areas that is different from those in midlatitude or temperate areas according to previous research

(Weng et al. 2009; Wu and Yuan 2007). According to the results, in tropical and subtropical regions, winter and spring are generally the better seasons to estimate impervious surfaces from optical remote sensing images compared with summer and autumn. Winter and spring are generally the dry seasons in tropical and subtropical regions and the temperature is relatively lower. With a specific investigation in Guangzhou, we found that in winter, there are not many clouds and most of the VSAs are not filled with water. Even though more bare soils in the VSAs are exposed, they can be easily identified because most are actually not dry soils as in the midlatitude areas. Therefore, satellite images are the most appropriate for estimating impervious surfaces. On the other hand, autumn images had the lowest accuracy of impervious surfaces due to the cloud coverage and water in VSAs. Autumn is a rainy season in a subtropical monsoon region, for which clouds occur very often and VSAs are always filled with water. Consequently, clouds are confused with bright impervious surfaces due to their similar high reflectance, and more water in VSAs is confused with dark impervious surfaces due to their similar low reflectance.

The seasonal sensitivity of the two methods was also compared. Both ANN and SVM methods showed general consistency in the accuracy of the seasonal changes. ANN was somewhat more stable as its accuracy changed less than that obtained using SVM. However, both methods indicated that wintertime is the best season for ISE with satellite images in subtropical monsoon regions. The limitations of this study mainly come from the methodology, which is generally based on a per-pixel level. In urban and suburban areas, one pixel with a size of 30 × 30 m does not necessarily include only impervious materials or nonimpervious materials (Weng 2012; Wu and Murray 2003). In this case, the use of per-pixel methods would obtain a result with lower accuracy.

8.1.2 Feature Extraction Methods

This book has proposed a novel feature extraction technique based on SAN to incorporate the advantages of human vision into the process of remote sensing images. Methodologies on how to determine the SAN of each pixel, how to extract the textural features and geometric features from each SAN, and how to integrate all these features, have been presented and analyzed in detail. Lastly, a set of experiments was designed to conduct the SAN feature extraction framework and applied to classify two study areas located in Hong Kong and Cape Town. Additional quantitative analysis was performed to evaluate related parameters to determine the SAN and to compare the influences of different features, such as the spectral feature, color feature, and SAN-integrated features, on the accuracy of classification. Some interesting results were found from the experiments. First, with various features extracted from the images, the accuracy could be improved compared with using only the original image data of optical images. This indicates the effectiveness of feature extraction for remote sensing classifications for LULC and ISE. Second, with the spectral, texture,and shape feature extraction, the combination of optical and SAR images obtained better results than by using optical data alone. This is consistent with the results in previous experiments, and also proves the effectiveness of the synergistic use of optical and SAR data. Third, edge effects located on the image edges can be found in all study cases in Chapters 6 and 7 due to the texture extraction based on the GLCM technique as it applies a moving window with a certain window size. The additional use of SAN texture and shape features reduced this edge effect to some extent. Moreover, the SAN shape feature was able to enhance the edge information on the boundaries between different land objects, which is the main reason why SAN was able to improve the classification accuracy in the Hong Kong study case. However, the SAN texture and shape feature did not necessarily improve the classification results because it may produce some noises depending on the land cover diversity of the study areas.

8.1.3 Comparison between Optical and SAR Data

This study compares optical and SAR data in terms of estimation of impervious surfaces using a single data source. Experimental results in four different cities of the tropical and subtropical regions show some important findings for both the advantages and disadvantages of each data source. First, in all the cases, using optical images alone provided a generally better result than using SAR data alone. The difference of overall accuracy varied from about 7% to about 29%, while the difference of the Kappa coefficient varied from about 11% to about 60%. The results demonstrate that using optical image alone provides generally better identification of vegetation, dark impervious surfaces, and bright impervious surfaces, even though there are spectral confusions between impervious surfaces and vegetation or bare soils. However, due to the speckle phenomenon of SAR images, the ISE results using SAR data alone were affected by numerous noises, especially on the boundaries between different land covers. These noises can influence the classification results dramatically and lower the accuracy depending on the complexity of land cover patterns. In particular, linear features such as roads and bridges cannot be correctly identified using only SAR data. Second, SAR data was able to show some advantages for ISE compared with optical data. For instance, the separation between bright impervious surfaces and bare soils could be reduced due to their different backscattering behaviors with microwave remote sensing. In addition, spectral confusions between dark impervious surfaces and vegetation could be reduced to some extent in the SAR images. Therefore, optical images and SAR images can provide complementary information for each other to improve the estimation of impervious surfaces. Third, by comparing the ANN and SVM classifiers, both methods demonstrated similar results when applied to the same dataset in the same study area. The difference of accuracy between the results from ANN and SVM are less than 1%. In general, our experimental results showed that SVM is more appropriate for using optical data alone, while ANN provided better results when using SAR data alone. However, this parameter is not so strong for all cases and it should depend on the land cover diversity in a specific application.

8.1.4 Fusion Level and Fusion Methods

This book has compared three different fusion levels, pixel level, feature level, and combinational level, regarding optical and SAR images in terms of ISE. Spectral and texture features were extracted from the optical and SAR images, as well as the detailed design of fusion strategies for pixel-, feature-, and combinational-level fusions. SVM was then employed to conduct the fusion operation. The experimental results showed some important conclusions for selecting the fusion strategy of fusing optical and SAR data. First, pixel-level fusion is not appropriate for optical and SAR image fusion, as it reduces the accuracy compared to the single use of optical data. This result was consistent in the four study cases. Second, both feature-level and combinational-level fusion are able to improve the accuracy of ISE by fusing optical and SAR data. However, whether feature-level or combinational-level fusion is better for the optical SAR fusion may depend on specific cases in terms of the land cover diversity and complexity. The resolutions of optical and SAR images may also influence the selection of an appropriate fusion strategy. Generally, the experiment demonstrated that combinational-level fusion is more suitable for ISE of urban areas with less diverse land covers, while feature-level fusion is more appropriate for urban areas with more diverse land covers.

8.2 Future Directions

8.2.1 Feature Extraction

Feature extraction is an important procedure for the processing of both optical and SAR images for accurate estimation of impervious surfaces, especially when high spatial resolution images are employed. The feature extraction approach based on SAN proposed in this book is efficient in extracting texture and shape features from high-resolution satellite images. However, the current research only applied it to the optical images. Texture features of SAR images are also very challenging according to the existing literature, and SAN-based feature extraction approach should also be applied to extract texture features from SAR images. However, due to the speckles in SAR images, the determination of SAN becomes especially difficult for SAR data. To address this problem, one possible solution is to use corresponding optical images (by coregistration) to determine the SAN, and then SAN-based texture features can be calculated on the SAR images.

8.2.2 Study Area Selection and Design

Six study sites, located in Guangzhou, Shenzhen, Hong Kong, Sao Paulo, Mumbai, and Cape Town, were selected for this research, and the corresponding Landsat TM, SPOT-5, ENVISAT ASAR, and TerraSAR-X images were used. Moreover, several experiments were designed, including the assessment of seasonal effects, the assessment of SAN-based feature extraction, the comparison between optical and SAR data, and the fusion of optical and SAR images. However, only one or two datasets were applied to each experiment due to the time limitation for preparing this research. Even though the results are consistent in all the experiments, from a statistical point of view, all datasets should be applied to every experiment for a more convincing result. Additionally, all the study sites selected in this research are from tropical and subtropical regions. In order to better understand the characteristics of ISE in tropical and subtropical urban areas, the most frequently studied regions, the temperate areas, should be compared to investigate the climatic and phenology influences, which is a topic for future research.

8.2.3 Validation with *In Situ* Data

Field data about various urban land covers was collected to help validate the visual interpretation in digital orthophoto. However, it is used only in a qualitative way. From the nature of this study, which focused on the combined use of optical and SAR data for the estimation of impervious surfaces, the current work is reasonable as the digital orthophoto is only used to help validate the visual interpretation for training and testing data selection. However, in future work, quantitative analysis will be required for the classification of digital orthophoto, which is important for subpixel analysis and assessment of classification results from satellite images.

8.2.4 Fusion Level and Strategy

Fusion level is an important factor when fusing multiple sources of satellite images. There are generally four different fusion levels for image fusion: the pixel level, the feature level, combinational level, and the decision level. In this book, the pixel level, feature level, and combinational level were compared. We found that pixel-level fusion is not appropriate for optical and SAR image fusion, as it reduces the accuracy compared to the single use of optical data. This result is consistent in the four study cases. Additionally, both feature-level and combinational-level fusion are able to improve the accuracy of ISE by fusing optical and SAR data. However, whether feature-level or combinational-level fusion is better for optical SAR fusion may depend on specific cases in terms of the land cover diversity and complexity. However, decision-level fusion was not implemented in this study since the decision rules remain very difficult to design for decision fusion. Nevertheless, the decision level, which is promising from examining the existing literature, should be investigated more by designing sophisticated decision rules in future studies. One possible solution for designing the decision rules is to simulate the reasoning procedure of human psychological cognition to take advantage of the great reasoning ability of human beings.

8.2.5 Fusion Methods

The fusion method is the conduction of fusion at a certain fusion level. Generally speaking, multisource data fusion refers to the combinational use of multiple data sources. That is to say, when we use multiple satellite images at the same time, we are fusing them. Therefore, whether a procedure can be called multisource data fusion depends not on what fusion methods are used, but on what data sources are used. Nevertheless, one challenge in multisource data fusion is how we fuse them, which means what fusion methods are used. In this book, three fusion methods are employed, including ANN, SVM, and RF. These methods have been applied to numerous application fields such as remote sensing classification, and they are known as the most popular nonparametric machine-learning approaches. However, from the view of multisource data fusion, these three methods have different behaviors when conducting the fusion procedure. For ANN, all the data sources are input into the input-layer nodes to determine the final results. That is, every data source is treated equally and they are used in the whole fusion procedure. For SVM, all the data sources are considered to determine the support vectors, which determine the final results. In this procedure, even though all data sources are treated equally when searching the support vectors, they are not necessarily selected as support vectors, and thus may not influence the final result equally. This is a good feature for the fusion of optical and SAR data since the two data sources are very different and they should not be treated in the same way (Soergel 2010). The third method is RF, which is similar to SVM in that all data sources are not necessarily used to determine the final results. A random selection of variables (from multiple sources) is conducted during the stage of building up the RF. The RF algorithm would evaluate the importance of each variable (e.g., Gini index) to determine whether it should be selected to form a tree node. In this way, each data source is treated differently in the RF method. This is actually the reason why RF obtained the best results in our study.

However, there are some challenging problems in the fusion of multiple satellite images, such as uncertainty. For instance, cloud coverage is a problem in optical images. When fusing two images of optical and SAR images with some clouds in the optical image, the information extracted from the optical image should be accompanied by some uncertainty depending on the occurrence and thickness of the clouds. However, this feature cannot be described in the fusion methods of ANN, SVM, and RF. Therefore, more advanced methods should be designed and applied to address this problem. One possible solution is to conduct decision fusion by designing decision rules to consider SAR data more than optical data in cloudy areas. Another possible way is to use incomplete data fusion theory and treat the cloudy areas as incomplete data by masking them out. Both of these approaches should be part of future research.

Appendix: Codes for Determining SAN

```
function [neighbor, peri] = SAN(img, g, i, j)
% This function is to calculate the SAN of a pixel at img(i,j)
% img is a false-color RGB image
% g is the color feature of img
% neighbor is the SAN of img(i,j) in one dimension
% peri is the perimeter of neighbor

global image              % a false-color RGB image
global N_vector           % one dimension SAN
global N                  % number of pixels in the SAN
global perimeter;         % perimeter of the SAN
global Nx;                %(Nx,Ny)is the size of image
global Ny;
global dd;                % distance between two pixels
global window_size;       % size of the viewport

image = img;
[Nx,Ny] = size(g);

N_vector = zeros(1,10000); % suppose the maximal SAN size is
                              10000
N = 0;
perimeter = 0;
dd = 100;                 % initialize of the distance
                            between two pixels
window_size = 11;         % define the size of viewport

centre = g(i,j);
setNeighbor(i,j,g,centre,i,j);

neighbor = N_vector(1:N);
peri = perimeter;

%%%%%%%%%%%%%%%%%%%%%%%%%%%%%%%%%%%%%%%%%%%%%%%%%%%%%%%%%%%%%%%%%%%
%%%%%%
function setNeighbor(i,j,g,centre,x,y)
global image;
global N_vector
global N
global perimeter
global Nx;
global Ny;
global dd;
```

```
global window_size;

bool = (i > (x-window_size)) && (j > (y-window_size)) && ...
       (i < (x+window_size)) && (j < (y+window_size)) ...
       && (i > 0) && (j > 0);
bool1 =  (image(i,j,1) == 255);
if(bool1 > 0)
      bool1 = 0;
else
      bool1 = 1;
end

if(i<Nx && j<Ny && bool && bool1)
      dd = abs(g(i,j)-centre);
end

if(dd<0.2 && i<Nx && j<Ny && bool && bool1)
      N = N + 1;
      N_vector(N) = image(i,j,1);

      image(i,j,1) = 255;
      image(i,j,2) = 0;
      image(i,j,3) = 0;

      setNeighbor(i+1,j,g,centre,x,y);
      if (i>2) setNeighbor(i-1,j,g,centre,x,y); end
      setNeighbor(i,j+1,g,centre,x,y);
      if (j>2) setNeighbor(i,j-1,g,centre,x,y); end
      setNeighbor(i+1,j+1,g,centre,x,y);
      if (i>2) setNeighbor(i-1,j+1,g,centre,x,y); end
      if (i>2 && j>2) setNeighbor(i-1,j-1,g,centre,x,y); end
      if (j>2) setNeighbor(i+1,j-1,g,centre,x,y); end
else
      perimeter = perimeter + 1;
end
```

Bibliography

Adams, J. B., D. E. Sabol, V. Kapos, R. Almeida, D. A. Roberts, M. O. Smith, and A. R. Gillespie. 1995. Classification of multispectral images based on fractions of endmembers—Application to land-cover change in the Brazilian Amazon. *Remote Sensing of Environment* 52 (2):137–154.

Arnold, C. L., P. J. Boison, and P. C. Patton. 1982. Sawmill Brook—An example of rapid geomorphic change related to urbanization. *Journal of Geology* 90 (2):155–166.

Arnold, C. L., and C. J. Gibbons. 1996. Impervious surface coverage—The emergence of a key environmental indicator. *Journal of the American Planning Association* 62 (2):243–258.

Bannerman, R. T., D. W. Owens, R. B. Dodds, and N. J. Hornewer. 1993. Sources of pollutants in Wisconsin stormwater. *Water Science and Technology* 28 (3–5):241–259.

Benz, U. C., P. Hofmann, G. Willhauck, I. Lingenfelder, and M. Heynen. 2004. Multiresolution, object-oriented fuzzy analysis of remote sensing data for GIS-ready information. *ISPRS Journal of Photogrammetry and Remote Sensing* 58 (3–4):239–258.

Bhagat, R. B. 2011. Emerging pattern of urbanization in India. *Economic & Political Weekly* 46:10–13.

Bharati, M. H., J. J. Liu, and J. F. MacGregor. 2004. Image texture analysis: Methods and comparisons. *Chemometrics and Intelligent Laboratory Systems* 72 (1):57–71.

Booth, D. B., and C. R. Jackson. 1997. Urbanization of aquatic systems—Degradation thresholds, stormwater detention, and limits of mitigation. *JAWRA Journal of the American Water Resources Association* 33 (5):1077–1090.

Breiman, L. 1984. *Classification and Regression Trees.* Belmont, CA: Wadsworth International Group.

Breiman, L. 2001. Random forests. *Machine Learning* 45 (1):5–32.

Breiman, L., J. H. Friedman, R. A. Olshen, and C. J. Stone. 1984. *Classification and Regression Trees.* Monterey, CA: Wadsworth.

Brinkhoff, Th. 2011. The principal agglomerations of the world—Population statistics & maps. citypopulation.de, accessed August 13, 2011.

Brun, S. E., and L. E. Band. 2000. Simulating runoff behavior in an urbanizing watershed. *Computers, Environment and Urban Systems* 24 (1):5–22.

Bruzzone, L., and D. F. Prieto. 2002. An adaptive semiparametric and context-based approach to unsupervised change detection in multitemporal remote-sensing images. *IEEE Transactions on Image Processing* 11 (4):452–466.

Cablk, M. E., and T. B. Minor. 2003. Detecting and discriminating impervious cover with high-resolution IKONOS data using principal component analysis and morphological operators. *International Journal of Remote Sensing* 24 (23):4627–4645.

Cai, W. J., M. H. Dai, Y. C. Wang, W. D. Zhai, T. Huang, S. T. Chen, F. Zhang, Z. Z. Chen, and Z. H. Wang. 2004. The biogeochemistry of inorganic carbon and nutrients in the Pearl River estuary and the adjacent Northern South China Sea. *Continental Shelf Research* 24 (12):1301–1319.

Calabresi, G. 1996. The use of ERS data for flood monitoring: An overall assessment. In *Second ERS Application Workshop*. London.

Chander, G., B. L. Markham, and J. A. Barsi. 2007. Revised Landsat-5 thematic mapper radiometric calibration. *IEEE Geoscience and Remote Sensing Letters* 4 (3):490–494.

Chander, G., B. L. Markham, and D. L. Helder. 2009. Summary of current radiometric calibration coefficients for Landsat MSS, TM, ETM+, and EO-1 ALI sensors. *Remote Sensing of Environment* 113 (5):893–903.

Chehata, N., L. Guo, and C. Mallet. 2009. Contribution of airborne full-waveform lidar and image data for urban scene classification. *2009 16th IEEE International Conference on Image Processing*, Vols. 1–6:1649–1652.

Civico, D. L., and J. D. Hurd. 1997. Impervious surface mapping for the state of Connecticut. Paper read at *Proceedings of ASPRS/ACSM Annual Convention*, April 7–10, Seattle, WA.

Clode, S., F. Rottensteiner, P. Kootsookos, and E. Zelniker. 2007. Detection and vectorization of roads from lidar data. *Photogrammetric Engineering and Remote Sensing* 73 (5):517–535.

Corbane, C., N. Baghdadi, X. Descombes, G. Wilson, N. Villeneuve, and M. Petit. 2009. Comparative study on the performance of multiparameter SAR data for operational urban areas extraction using textural features. *IEEE Geoscience and Remote Sensing Letters* 6 (4):728–732.

Cracknell, A. P. 1999. Twenty years of publication of the *International Journal of Remote Sensing*. *International Journal of Remote Sensing* 20 (18):3469–3484.

Curran, P. J. 1988. The semivariogram in remote-sensing—An introduction. *Remote Sensing of Environment* 24 (3):493–507.

Cutler, D. R., T. C. Edwards, K. H. Beard, A. Cutler, and K. T. Hess. 2007. Random forests for classification in ecology. *Ecology* 88 (11):2783–2792.

Cybenko, G. 1989. Approximation by superpositions of a sigmoidal function. *Mathematics of Control, Signals and Systems* 2 (4):303–314.

Dare, P. M. 2005. Shadow analysis in high-resolution satellite imagery of urban areas. *Photogrammetric Engineering and Remote Sensing* 71 (2):169–177.

Dekker, R. J. 2003. Texture analysis and classification of ERS SAR images for map updating of urban areas in the Netherlands. *IEEE Transactions on Geoscience and Remote Sensing* 41 (9):1950–1958.

Dell'Acqua, F., and P. Gamba. 2003. Texture-based characterization of urban environments on satellite SAR images. *IEEE Transactions on Geoscience and Remote Sensing* 41 (1):153–159.

Deng, C. B., and C. S. Wu. 2012. BCI: A biophysical composition index for remote sensing of urban environments. *Remote Sensing of Environment* 127:247–259.

Deng, C. B., and C. S. Wu. 2013. Examining the impacts of urban biophysical compositions on surface urban heat island: A spectral unmixing and thermal mixing approach. *Remote Sensing of Environment* 131:262–274.

Eastman, J. R. 2003. *IDRISI Kilimanjaro: Guide to GIS and Image Processing*. Manual Version 14. Worcester, MA: Clark University Press.

European Environment Agency (EEA). 2006. *EEA Fast Track Service Precursor on Land Monitoring—Degree of soil sealing 100m*, http://www.eea.europa.eu/data -and-maps/data/eea-fast-track-service-precursor-on-land-monitoring-degree -of-soil-sealing-100m-1.

Elberink, S. J. O., and G. Vosselman. 2009. 3D information extraction from laser point clouds covering complex road junctions. *Photogrammetric Record* 24 (125):23–36.

Elvidge, C. D., B. T. Tuttle, P. S. Sutton, K. E. Baugh, A. T. Howard, C. Milesi, B. L. Bhaduri, and R. Nemani. 2007. Global distribution and density of constructed impervious surfaces. *Sensors* 7 (9):1962–1979.

Espey, W. H., C. W. Morgan, and F. D. Masch. 1966. Study of some effects of urbanization on storm runoff from a small watershed. Texas Water Development Board, Report 23.

Fabbri, A. G., F. D. Vandermeer, C. R. Valenzuela, and C. A. Kushigbor. 1993. Shape-analysis and multispectral classification in geological remote-sensing. *Mathematical Geology* 25 (7):773–793.

Fan, F. L., Y. P. Wang, and Z. S. Wang. 2008. Temporal and spatial change detecting (1998–2003) and predicting of land use and land cover in Core corridor of Pearl River Delta (China) by using TM and ETM+ images. *Environmental Monitoring and Assessment* 137 (1–3):127–147.

Frankenberger, J. R., E. S. Brooks, M. T. Walter, M. F. Walter, and T. S. Steenhuis. 1999. A GIS-based variable source area hydrology model. *Hydrological Processes* 13 (6):805–822.

Galli, J. 1991. Thermal impacts associated with urbanization and stormwater management best management practices. In *Metropolitan Washington Council of Governments. Maryland Department of Environment.* Washington, DC.

Gamba, P., and M. Herold. 2009. *Global Mapping of Human Settlements: Experiences, Datasets, and Prospects.* Boca Raton, FL: CRC Press.

Gandy, M. 2008. Landscapes of disaster: Water, modernity, and urban fragmentation in Mumbai. *Environment and Planning A* 40 (1):108–130.

Germaine, K. A., and M. C. Hung. 2011. Delineation of impervious surface from multispectral imagery and lidar incorporating knowledge based expert system rules. *Photogrammetric Engineering and Remote Sensing* 77 (1):75–85.

Gislason, P. O., J. A. Benediktsson, and J. R. Sveinsson. 2006. Random forests for land cover classification. *Pattern Recognition Letters* 27 (4):294–300.

Goetz, S. J., R. K. Wright, A. J. Smith, E. Zinecker, and E. Schaub. 2003. IKONOS imagery for resource management: Tree cover, impervious surfaces, and riparian buffer analyses in the mid-Atlantic region. *Remote Sensing of Environment* 88 (1–2):195–208.

Graf, W. L. 1977. Network characteristics in suburbanizing streams. *Water Resources Research* 13 (2):459–463.

Guo, L., N. Chehata, C. Mallet, and S. Boukir. 2011. Relevance of airborne lidar and multispectral image data for urban scene classification using random forests. *ISPRS Journal of Photogrammetry and Remote Sensing* 66 (1):56–66.

Hagolle, O., M. Huc, D. V. Pascual, and G. Dedieu. 2010. A multi-temporal method for cloud detection, applied to FORMOSAT-2, VEN mu S, LANDSAT and SENTINEL-2 images. *Remote Sensing of Environment* 114 (8):1747–1755.

Hall, D. L., and J. Llinas. 1997. An introduction to multisensor data fusion. *Proceedings of the IEEE* 85 (1):6–23.

Ham, J., Y. C. Chen, M. M. Crawford, and J. Ghosh. 2005. Investigation of the random forest framework for classification of hyperspectral data. *IEEE Transactions on Geoscience and Remote Sensing* 43 (3):492–501.

Haralick, R. M., K. Shanmuga, and I. Dinstein. 1973. Textural features for image classification. *IEEE Transactions on Systems Man and Cybernetics* SMC3 (6):610–621.

Henderson, F. M., and Z. G. Xia. 1997. SAR applications in human settlement detection, population estimation and urban land use pattern analysis: A status report. *IEEE Transactions on Geoscience and Remote Sensing* 35 (1):79–85.

Herodotou, N., K. N. Plataniotis, and A. N. Venetsanopoulos. 1999. Automatic location and tracking of the facial region in color video sequences. *Signal Processing-Image Communication* 14 (5):359–388.

Hodgson, M. E., J. R. Jensen, J. A. Tullis, K. D. Riordan, and C. M. Archer. 2003. Synergistic use of lidar and color aerial photography for mapping urban parcel imperviousness. *Photogrammetric Engineering and Remote Sensing* 69 (9):973–980.

Hsieh, P. F., L. C. Lee, and N. Y. Chen. 2001. Effect of spatial resolution on classification errors of pure and mixed pixels in remote sensing. *IEEE Transactions on Geoscience and Remote Sensing* 39 (12):2657–2663.

Hsu, C., C. Chang, and C. Lin. 2007. *A Practical Guide to Support Vector Classification*. Taiwan: National Taiwan University, http://ntu.csie.org/~cjlin/papers/guide/guide.pdf.

Hu, X. 2011. Impervious surface estimation from remote sensing imagery using sub-pixel and object-based classifications in Indianapolis, USA. Department of Geography, Geology, and Anthropology, Indiana State University, Terre Haute, Indiana. ProQuest, UMI Dissertations Publishing, 134 pages, 2009. Dissertation/thesis number: 3394733.

Hu, X. F., and Q. H. Weng. 2009. Estimating impervious surfaces from medium spatial resolution imagery using the self-organizing map and multi-layer perceptron neural networks. *Remote Sensing of Environment* 113 (10):2089–2102.

Hu, X. F., and Q. H. Weng. 2011. Estimating impervious surfaces from medium spatial resolution imagery: A comparison between fuzzy classification and LSMA. *International Journal of Remote Sensing* 32 (20):5645–5663.

Huang, C., L. S. Davis, and J. R. G. Townshend. 2002. An assessment of support vector machines for land cover classification. *International Journal of Remote Sensing* 23 (4):725–749.

Hudson, I. L., S. W. Kim, and M. R. Keatley. 2009. Climatic influences on the flowering phenology of four eucalypts: A GAMLSS approach. *18th World IMACS Congress and Modsim09 International Congress on Modelling and Simulation*, 2611–2617.

Hurd, J. D., and D. L. Civco. 2004. Temporal characterization of impervious surfaces for the state of Connecticut. In *ASPRS Annual Conference Proceedings*. Denver, CO.

Inouye, D. W. 2008. Effects of climate change on phenology, frost damage, and floral abundance of montane wildflowers. *Ecology* 89 (2):353–362.

Jacobson, C. R. 2011. Identification and quantification of the hydrological impacts of imperviousness in urban catchments: A review. *Journal of Environmental Management* 92 (6):1438–1448.

Jat, M. K., P. K. Garg, and D. Khare. 2008. Monitoring and modelling of urban sprawl using remote sensing and GIS techniques. *International Journal of Applied Earth Observation and Geoinformation* 10 (1):26–43.

Jensen, J. R. 2007. *Introductory Digital Image Processing: A Remote Sensing Perspective* (Third Edition). London: Pearson Education Ltd.

Jensen, J. R., and D. C. Cowen. 1999. Remote sensing of urban suburban infrastructure and socio-economic attributes. *Photogrammetric Engineering and Remote Sensing* 65 (5):611–622.

Jiang, L. M., M. S. Liao, H. Lin, and L. M. Yang. 2009. Synergistic use of optical and InSAR data for urban impervious surface mapping: A case study in Hong Kong. *International Journal of Remote Sensing* 30 (11):2781–2796.

Jo, J. H., J. Carlson, J. S. Golden, and H. Bryan. 2010. Sustainable urban energy: Development of a mesoscale assessment model for solar reflective roof technologies. *Energy Policy* 38 (12):7951–7959.

Jones, G. V., and R. E. Davis. 2000. Climate influences on grapevine phenology, grape composition, and wine production and quality for Bordeaux, France. *American Journal of Enology and Viticulture* 51 (3):249–261.

Kavzoglu, T., and P. M. Mather. 2003. The use of backpropagating artificial neural networks in land cover classification. *International Journal of Remote Sensing* 24 (23):4907–4938.

Kikim, A., and P. S. Yadava. 2001. Phenology of tree species in subtropical forests of Manipur in north eastern India. *Tropical Ecology* 89 (2):353–362.

Klein, R. D. 1979. Urbanization and stream quality impairment. *Water Resources Bulletin* 15 (4):948–963.

Lee, D. H., K. M. Lee, and S. U. Lee. 2008. Fusion of lidar and imagery for reliable building extraction. *Photogrammetric Engineering and Remote Sensing* 74 (2):215–225.

Lee, J. S. 1983. Digital image smoothing and the sigma filter. *Computer Vision Graphics and Image Processing* 24 (2):255–269.

Lee, J. Y., and T. A. Warner. 2006. Segment based image classification. *International Journal of Remote Sensing* 27 (16):3403–3412.

Leinenkugel, P., T. Esch, and C. Kuenzer. 2011. Settlement detection and impervious surface estimation in the Mekong Delta using optical and SAR remote sensing data. *Remote Sensing of Environment* 115 (12):3007–3019.

Lippmann, R. 1987. An introduction to computing with neural nets. *IEEE ASSP Magazine* 4 (2):4–22.

Lombardo, P., M. Sciotti, T. M. Pellizzeri, and M. Meloni. 2003. Optimum model-based segmentation techniques for multifrequency polarimetric SAR images of urban areas. *IEEE Transactions on Geoscience and Remote Sensing* 41 (9):1959–1975.

Lopes, A., R. Touzi, and E. Nezry. 1990. Adaptive speckle filters and scene heterogeneity. *IEEE Transactions on Geoscience and Remote Sensing* 28 (6):992–1000.

Lu, D. S., S. Hetrick, and E. Moran. 2010. Land cover classification in a complex urban-rural landscape with QuickBird imagery. *Photogrammetric Engineering and Remote Sensing* 76 (10):1159–1168.

Lu, D. S., P. Mausel, M. Batistella, and E. Moran. 2004. Comparison of land-cover classification methods in the Brazilian Amazon Basin. *Photogrammetric Engineering and Remote Sensing* 70 (6):723–731.

Lu, D. S., E. Moran, and S. Hetrick. 2011. Detection of impervious surface change with multitemporal Landsat images in an urban-rural frontier. *ISPRS Journal of Photogrammetry and Remote Sensing* 66 (3):298–306.

Lu, D. S., H. Q. Tian, G. M. Zhou, and H. L. Ge. 2008. Regional mapping of human settlements in southeastern China with multisensor remotely sensed data. *Remote Sensing of Environment* 112 (9):3668–3679.

Lu, D. S., and Q. H. Weng. 2006. Use of impervious surface in urban land-use classification. *Remote Sensing of Environment* 102 (1–2):146–160.

Lu, D. S., and Q. H. Weng. 2009. Extraction of urban impervious surfaces from an IKONOS image. *International Journal of Remote Sensing* 30 (5):1297–1311.

Lu, D., and Q. Weng. 2007. A survey of image classification methods and techniques for improving classification performance. *International Journal of Remote Sensing* 28 (5):823–870.

Luo, L., and G. Mountrakis. 2010. Integrating intermediate inputs from partially classified images within a hybrid classification framework: An impervious surface estimation example. *Remote Sensing of Environment* 114 (6):1220–1229.

Marceau, D. J., P. J. Howarth, J. M. M. Dubois, and D. J. Gratton. 1990. Evaluation of the gray-level co-occurrence matrix-method for land-cover classification using SPOT imagery. *IEEE Transactions on Geoscience and Remote Sensing* 28 (4):513–519.

McFeeters, S. K. 1996. The use of the normalized difference water index (NDWI) in the delineation of open water features. *International Journal of Remote Sensing* 17 (7):1425–1432.

Miliaresis, G., and N. Kokkas. 2007. Segmentation and object-based classification for the extraction of the building class from LIDAR DEMs. *Computers & Geosciences* 33 (8):1076–1087.

Mingers, J. 1989. An empirical comparison of selection measures for decision-tree induction. *Machine Learning* 3:319–342.

Moghadam, H. S., and M. Helbich. 2013. Spatiotemporal urbanization processes in the megacity of Mumbai, India: A Markov chains-cellular automata urban growth model. *Applied Geography* 40:140–149.

Moller, M., L. Lymburner, and M. Volk. 2007. The comparison index: A tool for assessing the accuracy of image segmentation. *International Journal of Applied Earth Observation and Geoinformation* 9 (3):311–321.

Mountrakis, G., and L. Luo. 2011. Enhancing and replacing spectral information with intermediate structural inputs: A case study on impervious surface detection. *Remote Sensing of Environment* 115 (5):1162–1170.

Myint, S. W., C. P. Giri, W. Le, Z. L. Zhu, and S. C. Gillette. 2008. Identifying mangrove species and their surrounding land use and land cover classes using an object-oriented approach with a lacunarity spatial measure. *GIScience & Remote Sensing* 45 (2):188–208.

Myint, S. W., P. Gober, A. Brazel, S. Grossman-Clarke, and Q. H. Weng. 2011. Per-pixel vs. object-based classification of urban land cover extraction using high spatial resolution imagery. *Remote Sensing of Environment* 115 (5):1145–1161.

NOAA (National Oceanic and Atmospheric Administration). *Global Distribution and Density of Constructed Impervious Surfaces* 2010, http://www.ngdc.noaa.gov/dmsp /download _global_isa.html.

Pacifici, F., F. Del Frate, W. J. Emery, P. Gamba, and J. Chanussot. 2008. Urban mapping using coarse SAR and optical data: Outcome of the 2007 GRSS data fusion contest. *IEEE Geoscience and Remote Sensing Letters* 5 (3):331–335.

Pal, M. 2003. Random forests for land cover classification. *IGARSS 2003: IEEE International Geoscience and Remote Sensing Symposium, Proceedings*, Vols. I–VII: 3510–3512.

Pal, M. 2005. Random forest classifier for remote sensing classification. *International Journal of Remote Sensing* 26 (1):217–222.

Peel, M. C., B. L. Finlayson, and T. A. McMahon. 2007. Updated world map of the Koppen-Geiger climate classification. *Hydrology and Earth System Sciences* 11 (5):1633–1644.

Phinn, S., M. Stanford, P. Scarth, A. T. Murray, and P. T. Shyy. 2002. Monitoring the composition of urban environments based on the vegetation-impervious surface-soil (VIS) model by subpixel analysis techniques. *International Journal of Remote Sensing* 23 (20):4131–4153.

Puissant, A., J. Hirsch, and C. Weber. 2005. The utility of texture analysis to improve per-pixel classification for high to very high spatial resolution imagery. *International Journal of Remote Sensing* 26 (4):733–745.

Quattrochi, D. A., and M. F. Goodchild. 1997. *Scale in Remote Sensing and GIS.* New York: Lewis Publishers.

Quinlan, J. R. 1986. Induction of decision trees. *Machine Learning* 1:81–106.

Rashed, T., J. R. Weeks, and M. S. Gadalla. 2001. Revealing the anatomy of cities through spectral mixture analysis of multispectral satellite imagery: A case study of the greater Cairo region, Egypt. *Geocarto International* 16 (4):5–15.

Rebelo, A. G., P. M. Holmes, C. Dorse, and J. Wood. 2011. Impacts of urbanization in a biodiversity hotspot: Conservation challenges in Metropolitan Cape Town. *South African Journal of Botany* 77 (1):20–35.

Richards, J. A., and X. Jia. 2006. *Remote Sensing Digital Image Analysis: An Introduction* (4th Edition.) Berlin: Springer.

Ridd, M. K. 1995. Exploring a V-I-S (vegetation-impervious surface-soil) model for urban ecosystem analysis through remote-sensing—Comparative anatomy for cities. *International Journal of Remote Sensing* 16 (12):2165–2185.

Roberts, D. A., M. Gardner, R. Church, S. Ustin, G. Scheer, and R. O. Green. 1998. Mapping chaparral in the Santa Monica Mountains using multiple endmember spectral mixture models. *Remote Sensing of Environment* 65 (3):267–279.

Rokach, L. 2010. Ensemble-based classifiers. *Artificial Intelligence Review* 33 (1–2):1–39.

Rose, L. S., H. Akbari, and H. Taha. 2003. Characterizing the fabric of the urban environment: A case study of greater Houston, Texas. *LBNL-51448.* Berkeley, CA: Lawrence Berkeley National Laboratory.

Schneider, A., M. A. Friedl, and D. Potere. 2010. Mapping global urban areas using MODIS 500-m data: New methods and datasets based on "urban ecoregions." *Remote Sensing of Environment* 114 (8):1733–1746.

Schueler, T. 1994. The importance of imperviousness. *Watershed Protection Techniques* 1 (3):100–111.

Seabum, G. E. 1969. Effects of urban development on direct runoff to East Meadow Brook, Nassau County, Long Island, New York. *Professional Paper 627-B.* U.S. Geological Survey, Washington, DC.

Sheeder, S. A., J. D. Ross, and T. N. Carlson. 2002. Dual urban and rural hydrograph signals in three small watersheds. *Journal of the American Water Resources Association* 38 (4):1027–1040.

Shukla, R. P., and P. S. Ramakrishnan. 1982. Phenology of trees in a sub-tropical humid forest in northeastern India. *Vegetatio* 49 (2):103–109.

Sleavin, W. J., D. L. Civco, S. Prisloe, and L. Giannotti. 2000. Measuring impervious surfaces for non-point source pollution modeling. In *Proceedings of 2000 ASPRS Annual Convention.* Washington, DC.

Slonecker, E. T., D. B. Jennings, and D. Garofalo. 2001. Remote sensing of impervious surfaces: A review. *Remote Sensing Reviews* 20 (3):227–255.

Small, C. 2001. Estimation of urban vegetation abundance by spectral mixture analysis. *International Journal of Remote Sensing* 22 (7):1305–1334.

Small, C. 2002. Multitemporal analysis of urban reflectance. *Remote Sensing of Environment* 81 (2–3):427–442.

Soergel, U. 2010. *Radar Remote Sensing of Urban Areas.* Heidelberg: Springer.

Solberg, A. H. S., and A. K. Jain. 1997. Texture fusion and feature selection applied to SAR imagery. *IEEE Transactions on Geoscience and Remote Sensing* 35 (2):475–479.

Solso, R. L., M. K. MacLin, and O. H. MacLin. 2004. *Cognitive Psychology* (7th Edition). Cambridge, MA: Allyn & Bacon.

Stasolla, M., and P. Gamba. 2008. Spatial indexes for the extraction of formal and informal human settlements from high-resolution SAR images. *IEEE Journal of Selected Topics in Applied Earth Observations and Remote Sensing* 1 (2):98–106.

Staufer, P., and M. M. Fischer. 1997. Spectral pattern recognition by a two-layer perceptron: Effects of training set size. In *Neurocomputation in Remote Sensing Data Analysis* edited by I. Kanellopoulos, G. G. Wilkinson, F. Roli, and J. Austin, London: Springer, pp. 105–116.

Stumpf, A., and N. Kerle. 2011. Object-oriented mapping of landslides using random forests. *Remote Sensing of Environment* 115 (10):2564–2577.

Sun, Z. C., H. D. Guo, X. W. Li, L. L. Lu, and X. P. Du. 2011. Estimating urban impervious surfaces from Landsat-5 TM imagery using multilayer perceptron neural network and support vector machine. *Journal of Applied Remote Sensing* 5.

Tison, C., J. M. Nicolas, F. Tupin, and H. Maitre. 2004. A new statistical model for Markovian classification of urban areas in high-resolution SAR images. *IEEE Transactions on Geoscience and Remote Sensing* 42 (10):2046–2057.

Tiwari, P. S., H. Pande, and A. K. Pandey. 2009. Automatic urban road extraction using airborne laser scanning/altimetry and high resolution satellite data. *Journal of the Indian Society of Remote Sensing* 37 (2):223–231.

Torrens, P. M., and M. Alberti. 2000. Measuring sprawl. In *Working Paper No. 27, Centre for Advanced Spatial Analysis, University College.* London.

Torres, H., H. Alves, and M. A. De Oliveira. 2007. Sao Paulo peri-urban dynamics: Some social causes and environmental consequences. *Environment and Urbanization* 19 (1):207–223.

Tupin, F., and M. Roux. 2003. Detection of building outlines based on the fusion of SAR and optical features. *ISPRS Journal of Photogrammetry and Remote Sensing* 58 (1–2):71–82.

Tupin, F., and M. Roux. 2005. Markov random field on region adjacency graph for the fusion of SAR and optical data in radargrammetric applications. *IEEE Transactions on Geoscience and Remote Sensing* 43 (8):1920–1928.

United Nations. 2012. World urbanization prospects. The 2011 Revision. New York.

Vapnik, V. 1995. *The Nature of Statistical Learning Theory.* New York: Springer-Verlag.

Vapnik, V. 1998. *Statistical Learning Theory.* New York: Wiley.

Wang, L., and J. S. Cardenas. 2011. Small area population estimation with high-resolution remote sensing and lidar. In *Urban Remote Sensing: Monitoring, Synthesis and Modeling in the Urban Environment,* edited by X. Yang. Chichester, UK: John Wiley & Sons.

Waske, B., J. A. Benediktsson, and J. R. Sveinsson. 2009. Fusion of multisource data sets from agricultural areas for improved land cover classification. *2009 IEEE International Geoscience and Remote Sensing Symposium,* Vols. 1–5:3332–3335.

Waske, B., and S. van der Linden. 2008. Classifying multilevel imagery from SAR and optical sensors by decision fusion. *IEEE Transactions on Geoscience and Remote Sensing* 46 (5):1457–1466.

Weng, Q. H. 2001. Modeling urban growth effects on surface runoff with the integration of remote sensing and GIS. *Environmental Management* 28 (6):737–748.

Weng, Q. H. 2012. Remote sensing of impervious surfaces in the urban areas: Requirements, methods, and trends. *Remote Sensing of Environment* 117:34–49.

Weng, Q. H., and X. F. Hu. 2008. Medium spatial resolution satellite imagery for estimating and mapping urban impervious surfaces using LSMA and ANN. *IEEE
Transactions on Geoscience and Remote Sensing* 46 (8):2397–2406.

Weng, Q. H., X. F. Hu, and H. Liu. 2009. Estimating impervious surfaces using linear spectral mixture analysis with multitemporal ASTER images. *International
Journal of Remote Sensing* 30 (18):4807–4830.

Weng, Q. H., X. F. Hu, and D. S. Lu. 2008. Extracting impervious surfaces from
medium spatial resolution multispectral and hyperspectral imagery: A comparison. *International Journal of Remote Sensing* 29 (11):3209–3232.

Weng, Q. H., D. S. Lu, and B. Q. Liang. 2006. Urban surface biophysical descriptors and land surface temperature variations. *Photogrammetric Engineering and
Remote Sensing* 72 (11):1275–1286.

Weng, Q. H., D. S. Lu, and J. Schubring. 2004. Estimation of land surface temperature-
vegetation abundance relationship for urban heat island studies. *Remote
Sensing of Environment* 89 (4):467–483.

Weston, J., and C. Watkins. 1999. Support vector machines for multi-class pattern
recognition. Paper read at 7th European Symposium on Artificial Neural
Networks, April 21–23, Bruges, Belgium.

Wu, C. S. 2009. Quantifying high-resolution impervious surfaces using spectral mixture analysis. *International Journal of Remote Sensing* 30 (11):2915–2932.

Wu, C. S., and A. T. Murray. 2003. Estimating impervious surface distribution by
spectral mixture analysis. *Remote Sensing of Environment* 84 (4):493–505.

Wu, C. S., and F. Yuan. 2007. Seasonal sensitivity analysis of impervious surface estimation with satellite imagery. *Photogrammetric Engineering and Remote Sensing*
73 (12):1393–1401.

Wu, C. S., and A. T. Murray. 2005. A cokriging method for estimating population
density in urban areas. *Computers, Environment and Urban Systems* 29:558–579.

Xie, H., L. E. Pierce, and F. T. Ulaby. 2002. SAR speckle reduction using wavelet
denoising and Markov random field modeling. *IEEE Transactions on Geoscience
and Remote Sensing* 40 (10):2196–2212.

Xu, H. Q. 2006. Modification of normalised difference water index (NDWI) to enhance
open water features in remotely sensed imagery. *International Journal of Remote
Sensing* 27 (14):3025–3033.

Xu, H. Q. 2010. Analysis of impervious surface and its impact on urban heat environment using the normalized difference impervious surface index (NDISI).
Photogrammetric Engineering and Remote Sensing 76 (5):557–565.

Yang, L. M., C. Q. Huang, C. G. Homer, B. K. Wylie, and M. J. Coan. 2003a. An approach for
mapping large-area impervious surfaces: Synergistic use of Landsat-7 ETM+ and
high spatial resolution imagery. *Canadian Journal of Remote Sensing* 29 (2):230–240.

Yang, L. M., G. Xian, J. M. Klaver, and B. Deal. 2003b. Urban land-cover change detection through sub-pixel imperviousness mapping using remotely sensed data.
Photogrammetric Engineering and Remote Sensing 69 (9):1003–1010.

Yang, L. M., L. M. Jiang, H. Lin, and M. S. Liao. 2009a. Quantifying sub-pixel urban
impervious surface through fusion of optical and InSAR imagery. *GIScience &
Remote Sensing* 46 (2):161–171.

Yang, W., T. Y. Zou, D. X. Dai, and Y. M. Shuai. 2009b. Supervised land-cover classification of TerraSAR-X imagery over urban areas using extremely randomized
clustering forests. *2009 Joint Urban Remote Sensing Event*, Vols. 1–3:802–807.

Yang, X. G., W. Y. Fan, and Y. Yu. 2010. Estimation of forest canopy chlorophyll content based on PROSPECT and SAIL models. *Spectroscopy and Spectral Analysis* 30 (11):3022–3026.

Yu, D., and C. Wu. 2004. Understanding population segregation from Landsat ETM+ imagery: A geographically weighted regression approach. *GIScience & Remote Sensing* 41 (3):187–206.

Yu, X. W., J. Hyyppa, M. Vastaranta, M. Holopainen, and R. Viitala. 2011. Predicting individual tree attributes from airborne laser point clouds based on the random forests technique. *ISPRS Journal of Photogrammetry and Remote Sensing* 66 (1):28–37.

Yuan, F., and M. E. Bauer. 2007. Comparison of impervious surface area and normalized difference vegetation index as indicators of surface urban heat island effects in Landsat imagery. *Remote Sensing of Environment* 106 (3):375–386.

Zambon, M., R. Lawrence, A. Bunn, and S. Powell. 2006. Effect of alternative splitting rules on image processing using classification tree analysis. *Photogrammetric Engineering and Remote Sensing* 72 (1):25–30.

Zhang, H. S., H. Lin, Y. Li, and Y. Z. Zhang. 2013. Feature extraction for high-resolution imagery based on human visual perception. *International Journal of Remote Sensing* 34 (4):1146–1163.

Zhang, H., Y. Zhang, and H. Lin. 2012. A comparison study of impervious surfaces estimation using optical and SAR remote sensing images. *International Journal of Applied Earth Observation and Geoinformation* 18:148–156.

Zhang, J. X., J. H. Yang, Z. Zhao, H. T. Li, and Y. H. Zhang. 2010. Block-regression based fusion of optical and SAR imagery for feature enhancement. *International Journal of Remote Sensing* 31 (9):2325–2345.

Zhang, Y. 2001. Texture-integrated classification of urban treed areas in high-resolution color-infrared imagery. *Photogrammetric Engineering and Remote Sensing* 67 (12):1359–1365.

Zhang, Y. H., M. Hu, L. J. Zhong, A. Wiedensohler, S. C. Liu, M. O. Andreae, W. Wang, and S. J. Fan. 2008. Regional integrated experiments on air quality over Pearl River Delta 2004 (PRIDE-PRD2004): Overview. *Atmospheric Environment* 42 (25):6157–6173.

Index

Page numbers ending in "f" refer to figures. Page numbers ending in "t" refer to tables.

Printed and bound by CPI Group (UK) Ltd, Croydon, CR0 4YY

01/11/2024

01782619-0018